Learn to Produce Videos with FFmpeg: In 30 Minutes or Less

By Jan Ozer

Doceo Publishing, Inc.

Learn to Produce Videos with FFmpeg: In Thirty Minutes or Less

Jan Ozer

Doceo Publishing
412 West Stuart Drive
Galax, VA 24333

www.doceopub.com
www.streaminglearningcenter.com

ISBN: 978-0-9984530-1-9
Printed in the United States of America

To my daughters,

Elizabeth and Eleanor

*whose efforts and
accomplishments inspire me*

Acknowledgments

This book started life as the FFmpeg-related components of *Video Encoding by the Numbers*, and later morphed into a presentation at Streaming Media West which I hosted along with David Hassoun, with major assistance from Jun Heider, both from premiere consultancy RealEyes (http://realeyes.com/). My thanks to them and to the Streaming Media team of editorial and trade show staffers. Love you guys.

Thanks to HP for providing the encoding platforms I use in my testing and work. This and the *Video Encoding* book required thousands of encodes and other computations; I couldn't have done it without my HP Z840 and other HP computers and notebooks.

This book is the ninth published by my company, Doceo Publishing. As always, budgets are tight, time is short, and the topics are fast moving, so I apologize in advance for any rough edges. We'll try to do better next time.

As always, thanks to Pat Tracy for technical and marketing assistance.

Contents

Contents

Introduction

FFmpeg is a command line tool used to process video. In this book, you'll learn how to install FFmpeg, create command line arguments for encoding video with FFmpeg, and how to create files you can use to distribute video over the Internet. Let's start with a brief look at my assumptions about you, my theory for writing this book, and what's in it for you.

About You

Here's what I assume about you. I share these because they guided what I included in the book, and what I left out.

- You may know very little about video. You can catch up in a primer available in Chapter 1, Video Boot Camp. There will also be mini-sessions about the parameters we're discussing in each chapter.

- You know a bit about batch file creation and operation, but may not be an expert. I'll either teach you what you need to know, or point you to better sources.

My Theory for This Book

I recently completed a book called *Video Encoding by the Numbers*, which presents objective evidence for all recommended encoding decisions. It's completely research driven and involved thousands of test encodings.

To produce those encodings, I decided to learn to use FFmpeg. Since I was learning more and more about FFmpeg, I decided to include the techniques used for each chapter in that chapter of that book. The content turned out to be so useful that several readers suggested creating a separate book just for FFmpeg production.

So, that's what this book is. Specifically, this book contains some of the high-level findings that I presented in *Video Encoding* without much of the research, explanation, and background material that comprises the primary value in that book. This book also contains a lot more information on FFmpeg installation and operation.

If your primary interest is learning and using FFmpeg, there's no reason to buy *Video Encoding*. But if you want to learn the theory behind all the encoding and distribution decisions, you probably will find *Video Encoding* useful as a totally separate resource.

With all that said, here's my theory for this book.

- I'm knowledgeable about video, so I'll try and teach you the basics you need to know to understand how to produce streaming video.

- I'm pretty knowledgeable about encoding video with FFmpeg and I'll teach you what you need to know in this book. This book is a sharp-edged tool, however, and focuses on file conversions and encoding for streaming only, not the thousands of other operations that FFmpeg can perform. If you're looking for a general-purpose reference, check out Frantisek Korbel's *FFmpeg Basics: Multimedia Handling with a Fast Audio and Video Encoder*. I bought a copy for my first FFmpeg project, and it's here by my side as I type.

- I'm pretty competent with Windows batch file operation, been doing it for decades. So I'll teach you what you need to know here as well.

- I don't know much at all about Mac/Linux command line processing, so I'll point you towards other resources for the nitty gritty details about installation and operation. I have verified the operation of all batch files presented in this book in Windows, but not Mac and Linux. They should work as is, but you may need to make some minor tweaks.

- Overall, my guiding principle is to not reinvent the wheel, particularly as it relates to ancillary matters like installation, batch file creation, or the like. If there's a perfectly good tutorial or explanation out there, I'll point you to it, and together we'll save some trees.

What You'll Get From This Book

In this book, you will learn:

- the video basics needed to understand fundamental streaming production

- how to install FFmpeg on Windows, Mac, and Linux computers

- how to create batch files on all three operating systems

- how to encode files for streaming with FFmpeg, and some ancillary operations

- a bit about the best encoding choices for each parameter that we discuss.

When you buy a PDF version of this book, you'll also get a zip file containing access to all the Windows batch files and input and output files referenced in the book so you can use these for further testing or to accelerate your own projects. If you buy a print version of the book, please e-mail a copy of your receipt to ffmpeg@streaminglearningcenter.com and we'll send you the materials.

Enough talking about it, let's get going.

Chapter 1: Video Boot Camp

Figure 1-1. Too much compression makes your videos look ugly.

This book is targeted towards readers who may not know that much about video and streaming. So, in this chapter, you'll learn the video and streaming-related terms and concepts necessary to understand the more technical conversations to follow. Specifically, you will learn:

- about compression and codecs

- how to choose the right codec

- what a container format is and why you care

- the difference between progressive download, streaming, and adaptive streaming

- basic file parameters, like resolution, frame rate, and data rate

- some basics about video quality metrics like PSNR

- some background on the tables in this book.

Since the concept of compression is absolutely pervasive to all streaming media, we'll start with a quick look at the definition of codecs and compression.

Compression and Codecs

Compression technologies shrink your audio/video streams down to sizes that you can deliver to desktop and mobile viewers over the Internet. Compression is also the technology that can make your video ugly when you apply too much of it—as Figure 1-1 shows.

That's because all video compression technologies are lossy, which means they throw away information during compression. Upon decompression, lossy technologies create only an approximation of the original frame, not an exact replica. The more you compress, the more information gets thrown away and the worse the approximation looks. Obviously, we threw away too much data in the frame shown on the right in Figure 1-1.

What's a Codec?

Codecs are compression technologies with two components: an enCOder to compress the file in your studio or office and a DECoder to decode the file when played by the remote viewer. As this nifty capitalization suggests, codec is a contraction of the terms "encoder" and "decoder."

There are many video codecs—like H.264, H.265, MPEG-4, VP8, VP9, MPEG-2 and MPEG-1—and lots of audio codecs—like MP3 and Advanced Audio Coding (AAC). Table 1-1 shows the video codecs we'll be working with in this book; H.264/AVC, H.265/HEVC, and VP9.

Codec	H.264/AVC	H.265/HEVC	VP9
Date First Available	2003	2013	2013
Originator	ISO/MPEG	IOS/MPEG	Google
Compression Efficiency	The baseline	2x H.264	2x H.264
Typical Resolution	Up to 1080p	Up to 4K	Up to 4K
Typical Use	Everywhere	Smart TVs and STBs	Browser/YouTube

Table 1-1. The video codecs you will work with in this book.

Here's a brief introduction to the video codecs that you'll work with in this book.

- **H.264 (also called AVC).** Today's "it codec," H.264 plays almost everywhere, so is as close to a universal video codec as is available today. H.264 is typically used at up to 1080p resolution, although it can extend beyond 1080p. You'll about H.264-specific encoding options in Chapter 8.

- **HEVC (High Efficiency Video Coding, also called H.265).** H.265/HEVC is the standards-based successor to H.264/AVC that plays primarily on 4K TVs—although Android 5 includes a software decoder and recent iPhones use HEVC for FaceTime. HEVC is not used for general-purpose streaming (at least in 2017) because playback isn't available in any browsers. H.265 is roughly twice as efficient as H.264, which means the same video quality at about 50 percent the data rate of H.264. You'll learn about HEVC in Chapter 12.

- **VP9.** Google's open-source competitor to H.265, which delivers about the same quality as HEVC. Since playback is available in every browser but Safari, it's much more accessible than HEVC for streaming usage. You'll learn about VP9 in Chapter 13.

Several other codecs are worth mentioning for completeness's sake. MPEG-2 is an older standards-based codec created for broadcast and DVD/Blu-ray discs. It's still very widely used in broadcast applications, but never for streaming. MPEG-4 is standards-based video codec about 25 percent less efficient than H.264 that has been supplanted by H.264 except for very old devices.

On2's VP6 was the codec that powered Adobe Flash until it was supplanted by H.264 in 2006. Ogg Theora was the codec that powered HTML5 video until Google launched VP8, which made Ogg disappear overnight like a magic trick (never did like Ogg). Looking ahead, the AV1 codec from the Alliance for Open Media should debut by the end of 2017, and it will slowly replace VP9 and HEVC in many applications.

Audio Codecs

Most of your video files will have audio components which must be compressed as well. Here are the audio codecs you'll work with in this book.

- *Advanced Audio Coding (AAC)*. AAC audio codecs are used with the H.264, H.265, and MPEG-4 video codecs.

- *Opus codec.* Opus is the audio codec used with VP8 and VP9.

Choosing a Codec

Though you may not be in charge of choosing a codec, some background on where and when to use them certainly can't hurt. Here are the considerations involved in choosing a codec.

First, when you target a particular distribution platform like computers or mobile, you must make sure that platform includes the ability to decode the files you're about to send them. For example, iPhones play H.264 video, but not MPEG-2 or H.265 (at least not in early 2017). If you want video to play on an iPhone, encode it using the H.264 codec. In fact, H.264 is as close to a universal codec as we have in this world. It plays in Flash, in Silverlight, in most browsers, on all mobile and over the top (OTT) platforms like Roku, on Apple TV, and on all smart TVs.

Beyond compatibility, you should also consider suitability. For example, most smart TVs can play H.264 and H.265, but H.265 is more efficient than H.264. So, Netflix, Hulu, Amazon, and other companies targeting these TV sets use H.265. Most of these sets also play VP9 and will play AV1 when available, so I expect many services to switch to VP9 and later to AV1.

Container Formats

Another fundamental decision you make when encoding a file is the container format, which is likely a familiar concept. Let's start with an easy one. If you see a file with a .mov extension, you probably think QuickTime file, which means that the data within the file is stored in a way that the QuickTime Player, and products built around the QuickTime standard, understand. In these cases, the video in the file is the content, and QuickTime is the container format.

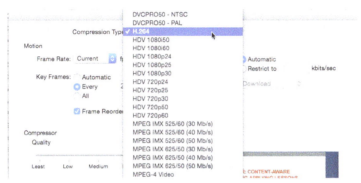

Figure 1-2. You can output many different codecs in a QuickTime file with a .mov extension.

What's critical to understand is that a container format is separate from the codec. You can see this in Figure 1-2, the standard video compression settings screen from QuickTime Pro. While you can create the file with all the compression technologies shown, each file would have a .mov extension. The codec is how the video is compressed, while the container format controls how that compressed data is stored within the file.

Things can get confusing when you mix codecs and container formats. For example, HTTP Live Streaming (HLS), a technology used to distribute video to Apple devices and other endpoints, uses H.264 encoded video stored in an MPEG-2 transport stream (MPEG-TS) container format. When you encode with FFmpeg, you can set both codec and container format; so long as you remember that they are separate concepts, you should have no trouble sorting this out.

Distribution Alternatives

When deploying web video technologies, it's important to recognize that they offer varying delivery options, including progressive download, streaming, and adaptive streaming. Since these techniques are critical to streaming operation, let's describe how they work.

- **Progressive download.** Video delivered by a standard HTTP web server, just like any other form of content. Video delivered via progressive download is typically delivered as fast as the web server can send it, which wastes bandwidth if the viewer stops watching after significant portions of the file have been delivered. Because of this inefficiency, progressive download can also degrade service when delivering to multiple viewers.

- **Streaming.** Video delivered by a streaming server. A streaming server meters out video as it's consumed by the player, so bandwidth isn't wasted if the viewer stops watching. This schema also allows a server to deliver more efficiently to multiple viewers.

- *Adaptive streaming.* An advanced schema for serving different platforms over disparate connections. Involves creating multiple iterations from the same live or on-demand file at different bitrates and resolutions and adaptively switching streams based upon the viewer's connection speed and playback device. You can see an example of the files created for adaptive streaming in Table 1-2. Technologies that enable adaptive streaming are often called adaptive bitrate technologies and abbreviated ABR.

Configuration Basics

Apple's adaptive streaming technology, called HTTP Live Streaming (HLS), dominates mobile and over-the-top (OTT) platforms. Apple provides specific encoding recommendations for HLS in a document entitled, "HLS Authoring Specification for Apple Devices," which you can find at bit.ly/A_Devices_Spec. Table 1-2 is a snippet from that document.

Video average bit rate (kb/s)	Resolution	Frame rate
145	416 x 234	≤ 30 fps
365	480 x 270	≤ 30 fps
730	640 x 360	≤ 30 fps
1100	768 x 432	≤ 30 fps
2000	960 x 540	same as source
3000	1280 x 720	same as source
4500	same as source	same as source
6000	same as source	same as source
7800	same as source	same as source

Table 1-2. Encoding recommendations from Apple.

The Apple spec details file parameters like resolution, frame rate, and bitrate (also called data rate). That's because these three parameters are absolutely essential to video quality. Mess up any one of these, and quality will be irreparably compromised. Get them right, and you have a foundation for excellent quality.

Understanding what these configuration options are and how they interrelate will allow you to make better, more informed decisions, and avoid some potholes. So, let's start with the first configuration item in the Apple document, resolution.

Tip: The collection of files in Table 1-2 is referred to as an encoding ladder, and the individual configurations are often called rungs on the ladder, or variants when working with HLS. When encoding for adaptive streaming, selecting and configuring the individual rungs in the ladder is a critical step.

Video Resolution

Resolution is the width and height of the video in pixels (Figure 1-3). Most video is originally captured at 720x480 (NTSC Standard Definition) or 720x576 (PAL standard definition); 1280x720 or 1920x1080 (high-definition); or 3840x2160 or higher (ultra-high definition).

Figure 1-3. Resolution is the width and height of pixels in the file.

However, often these high-resolution files get scaled to smaller resolutions for streaming. This scaling reduces the number of pixels being encoded, making the file easier to compress while retaining good quality. For example, a 640x360 video has 230,400 pixels in each frame, while a 1280x720 video file has 921,600 pixels—or four times as many, as shown in Figure 1-4.

Referring to Table 1-2, this is why Apple recommends dropping to lower resolutions as bitrates decrease. While you're losing resolution and some detail, you're avoiding the kind of gross blockiness and other artifacts that mar overly compressed video files, like that shown on the right in Figure 1-1.

Note: Resolutions are often abbreviated based on the height, or second value. So 1920x1080 resolution is called 1080p, 1280x720 is called 720p, 960x540 is called 540p, 848x480 is 480p, 640x360 is called 360p, and so on. The p stands for progressive, which means the frame is a complete frame, rather than two fields as it used to be with interlaced video, which is now almost never shot or even seen. Sometimes frame rate (covered next) is thrown into the abbreviation, so 720p24 would be 1280x720 at 24 progressive frames per second.

Figure 1-4. Scaling a file to a lower resolution reduces the number of pixels.

Frame Rate

Most video starts life at 29.97 or 24 frames per second (fps), or 25 fps in Europe. Usually, producers who shoot at 24 fps deliver at that rate, while some producers who shoot at 29.97 fps deliver at 15 fps or even 10 fps when distributing to devices with a very slow Internet connection. That's because dropping the frame rate by 50 or 66 percent reduces the number of pixels being encoded, just like dropping the resolution from 1280x720 to 640x360.

You can see in Table 1-2 that Apple recommends dropping the frame rate on the first four files. That's because losing a bit of playback smoothness is preferable to blockiness and other gross compression artifacts.

Bitrate (or Data Rate)

Bitrate (or data rate) is the amount of data per second in the encoded video file, usually expressed in kilobits per second (kbps) or megabits per second (Mbps). In Table 1-2, Apple recommends a video bitrate of 145 kbps for the lowest-quality file, increasing to 7800 kbps for the highest-quality file.

By now you're seeing a relationship between bitrate, resolution, and, to a lesser degree, frame rate. At lower bitrates, which are necessary to deliver to devices with slower connection speeds, Apple recommends lower resolutions and frame rates, which limits the number of pixels being encoded. This makes it easier for the encoder to produce a high-quality file without blockiness or other artifacts.

Compression and Ben and Jerry's Ice Cream

Now that you know the basics, the easiest way to think about video compression is to compare it to a can of paint. Imagine a small can, like the size of a pint of Ben and Jerry's ice cream. If all you're painting is your front door, you're probably in good shape. On the other hand, if you're trying to paint the entire back wall of your house with that can, you're going to run into trouble. While you could spread the paint over the entire wall, the coverage would be sketchy, with lots of blotches of old paint showing through. To make it look really good, you just need more paint.

So it is with video compression. The size of the paint can is the bitrate per second in the file. The size of the wall you have to paint are the pixels per second in the file. Anything that increases the number of pixels per second—whether it's a larger resolution or more frames per second—increases the size of the wall you have to paint. At some point, the can is simply too small, and your video will look ugly. The only solutions are to add more paint and encode at a higher bitrate, or to decrease the size of the wall by encoding at a lower resolution or frame rate.

Returning to video compression, while there are some tricks you can play to make video look good at lower bitrates, if you see artifacts like those in Figure 1-1, it's likely because your data rate is too low for the resolution and frame rate you've selected. If so, you can either increase your data rate, or decrease your resolution or frame rate to improve the quality of your video.

About Video Quality Metrics

Part of what I tried to do in this book is to teach you the why as well as the how. Not as much as I did in *Encoding by the Numbers*, but enough to provide general guidance. As an example, in addition to detailing how to set the keyframe interval, I show you how that decision impacts quality, usually via data like that displayed in Table 1-3.

To create Table 1-3, I encoded eight test files to identical configurations and varied the keyframe interval (Chapter 7) from one every half of a second to one every ten seconds. The values shown in the table are Peak Signal-to-Noise Ratio (PSNR) values. To compute this, the tool I use, the Moscow State University Video Quality Measurement Tool, compares the compressed file to the source and measures the error in the compressed file. Higher scores indicate better quality, and the highest score for each row is in green, the lowest in red. This color schema allows you to look at the table and instantly realize that longer keyframe intervals produce better quality.

	.5 Sec	1 Sec	2 Sec	3 Sec	5 Sec	10 Sec	Max Delta
Tears of Steel	38.22	39.05	39.49	39.64	39.74	39.87	4.32%
Sintel	37.09	38.06	38.57	38.75	38.97	39.08	5.37%
Big Buck Bunny	37.03	37.93	38.52	38.68	38.64	39.09	5.57%
Talking Head	43.63	44.10	44.40	44.51	44.61	44.68	2.42%
Freedom	40.33	40.67	40.88	40.96	40.99	41.03	1.72%
Haunted	41.89	42.20	42.35	42.39	42.45	42.49	1.44%
Average	39.26	39.96	40.37	40.51	40.59	40.75	3.88%
Screencam	35.35	38.13	37.68	38.86	40.78	41.26	16.71%
Tutorial	38.26	43.06	43.61	44.65	46.15	47.89	25.17%

Table 1-3. PSNR values for different keyframe settings.

Why did I use eight different files? Because different types of content encode differently and I wanted a good cross section of animated and real world footage, as well as business-oriented footage like screencams and tutorials created out of PowerPoint slides and videos. Here's a description of the files that I tested.

- **Tears of Steel.** A Blender Foundation movie—mostly real-world video with some animation

- **Sintel.** Another Blender Foundation movie—all animation, but very lifelike rather than cartoonish

- **Big Buck Bunny.** Yet another Blender Foundation movie—all animation, but more cartoonish than Sintel

- **Screencam.** A demo video you can watch at bit.ly/vqmt_demo

- **Tutorial.** A PowerPoint presentation with talking head video grabbed from a Udemy course called "Encoding for Multiple Screen Delivery" (bit.ly/tut_vid)

- **Talking Head.** A simple talking-head video of yours truly in my office

- **Freedom.** Multicam concert footage (HDV/AVCHD) of the fabulous Josiah Weaver at the Greensboro Coliseum (vimeo.com/6044024)

- **Haunted.** Footage from a trailer I shot with a DSLR for the Haunted Graham Mansion (bit.ly/haunted_graham)

In *Encoding by the Numbers*, I performed and described many experiments detailing different approaches to encoding each kind of video. This book contains a subset of that data.

Now that you've got the video basics down, let's get FFmpeg installed and running, and learn some fundamentals of batch operation.

Chapter 2: Installing FFmpeg and Batch File Operation

Figure 2-1. The FFmpeg logo.

In the last chapter, you learned video basics, in this chapter, you'll learn batch file basics, as well as how install FFmpeg. Specifically, you will learn:

- what FFmpeg is

- how to install FFmpeg on Windows, Mac, and Linux computers

- what a batch file is and how to create one

- navigational fundamentals for all three operating systems

- how to run batch files on Windows, Mac, and Linux computers.

Note that the batch files shown in this chapter are for illustrative purpose and are not included in the ancillary materials.

About FFmpeg

Here's the skinny from Wikipedia:

FFmpeg is a free software project that produces libraries and programs for handling multimedia data. FFmpeg includes libavcodec, an audio/video codec library used by several other projects, libavformat (Lavf), an audio/video container mux and demux library, and the ffmpeg command line program for transcoding multimedia files. FFmpeg is published under the GNU Lesser General Public License 2.1+ or GNU General Public License 2+ (depending on which options are enabled).

The name of the project is inspired by the MPEG video standards group, together with "FF" for "fast forward". The logo uses a zigzag pattern that shows how MPEG video codecs handle entropy encoding.

So FFmpeg really stands for fast forward MPEG. Who knew? The key points, of course, are that FFmpeg is available for free (donations gladly accepted) and runs on Linux, Windows, and the Mac (Figure 2-2). Also, FFmpeg is the basis of homegrown encoding farms created by behemoths like YouTube and Netflix, as well as cloud encoding services like encoding.com or Hybrik. So, it's the real deal, in many ways the gold standard for encoding technologies.

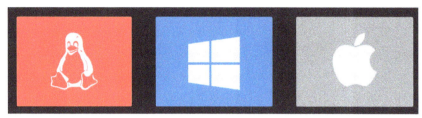

Figure 2-2. Packages available for Linux, Windows, and Mac at *https://ffmpeg.org/download.html*.

Installing FFmpeg

OK, let's get FFmpeg installed on your computers.

Installing on Ubuntu

I am not an expert on Linux. The Linux-based computer I'm working with runs Ubuntu 16.04 LTS, which comes with FFmpeg pre-installed. Boom, we're done. If you're running earlier versions of Ubuntu that don't come with FFmpeg installed, like 14.04, and you can't upgrade, check out the article entitled, How to Install FFmpeg in Ubuntu 14.04, 14.10 and Linux Mint, at bit.ly/FFmpeg_u_old.

Once installed, you should be able to run FFmpeg from any disk location without inserting the specific address for FFmpeg in the command line.

Installing on Windows

Installing on Windows is a two-part process; first, you download and unzip, then you insert FFmpeg into your Environmental variables so you can run it from any location without pointing back to the actual file location. In your batch files, this means you can use

```
ffmpeg
```

rather than

```
"c:\program files\ffmpeg\bin\ffmpeg"
```

or something similar. This definitely makes the five-minute process well worth your while.

The best tutorial I've seen for completing both steps is entitled, How to Install FFmpeg on Windows: 10 Steps (with Pictures) on Wikihow. You can find the article at bit.ly/FFmpeg_win.

> **Tip:** *Whenever there are spaces in a command line argument, like between program and files in the line just above, putting quotation marks at the start and end tells the operating system to treat the entire text string as a command, and not to break it up at the space. Without the quotes above, FFmpeg wouldn't run, because the operating system would read* `c:\program,` *stop at the space, and not know what to do.*

Installing on the Mac

Installing on the Mac is a three-step process;

1. **Install Xcode from the App store.** This is a free development environment for macOS, iOS, watchOS and tvOS.

2. **Install Homebrew.** In the words of the Homebrew website, "Homebrew installs the stuff you need that Apple didn't." To install Homebrew, you have to open a Terminal session and type this command at the prompt.

```
/usr/bin/ruby -e "$(curl -fsSL https://raw.githubusercontent.com/Homebrew/
install/master/install)"
```

You can more easily copy and paste this into Terminal from the Homebrew website, which is at https://brew.sh/. Briefly, Terminal is a utility that ships with all Macs. It's located in the Utilities folder off the Applications folder. You can either find and click it there, or search for it using Spotlight Search.

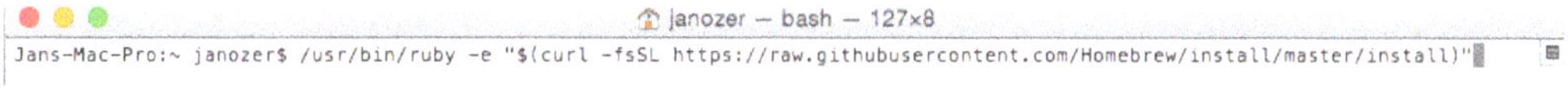

Figure 2-3. Mac Terminal installing Homebrew.

Learning how to access and use Terminal will be absolutely essential to running FFmpeg, so if you're not familiar with the application, search for and review some resources on the topic.

3. **Install FFmpeg**. Once you install Brew, you basically tell it to go find FFmpeg and install it on your system. To accomplish this, type this command into Terminal.

```
brew install ffmpeg
```

The hard part is telling Brew which options to install, which you do by using different switches. To install all available options, use this command from bit.ly/install_ff_mac.

```
brew install ffmpeg --with-fdk-aac --with-ffplay --with-freetype --with-
frei0r --with-libass --with-libvo-aacenc --with-libvorbis --with-libvpx
--with-opencore-amr --with-openjpeg --with-opus --with-rtmpdump --with-
schroedinger --with-speex --with-theora --with-tools
```

This is the one I used, and it worked just fine. For information about compiling FFmpeg yourself, check out the FFmpeg Mac Compilation Guide at bit.ly/Mac_Comp_FF. As with Linux (which, like the MacOS, is based on UNIX), once installed, FFmpeg should be available from any disk location without pointing back to the actual FFmpeg.exe program.

Working with Batch Files

Once you have FFmpeg installed, you can run it from the command line (Windows) or Terminal (Mac and Linux). For example, consider batch 2-1.

```
ffmpeg  -i ZOO_1080p.mov -c:v libx264 ZOO_1.mp4
```

Batch 2-1. A simple FFmpeg command line argument.

As you'll learn, this runs FFmpeg and tells it to encode the input file, ZOO_1080p.mov which is designated by the -i switch, using the x264 codec, and to create a file named ZOO_1.mp4. Easy, peasy. Simple enough to do if you only have one file to encode, and that's what I've done in Figure 2-4 (see the top line at the J:\FFMPEG> command prompt.

Figure 2-4. Encoding a single file in the Windows command line.

But what do you if you have multiple files to encode and don't want to feed them to FFmpeg one at a time? That's when you build and execute a batch file. In essence, a batch file is a series of lines of text to be executed serially, one after another, just as if you had typed them into the Terminal window yourself.

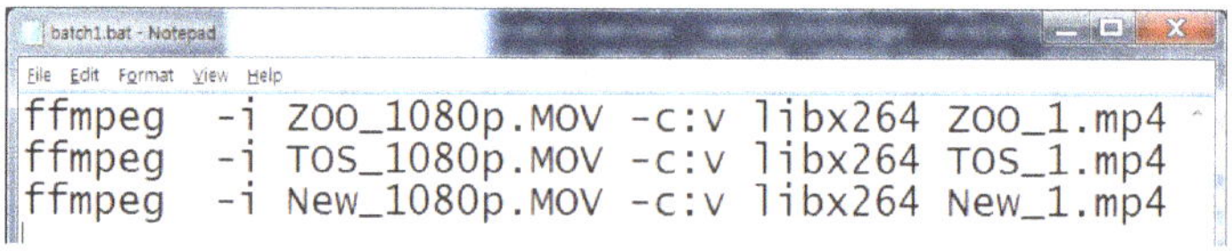

Figure 2-5. A simple Windows batch file.

Figure 2-5 is a Windows batch file named batch1.bat (.bat designates the file as a batch file in Windows). In an FFmpeg batch file, you always list FFmpeg first, which tells the OS to run FFmpeg. Then you include a bunch of "switches" or "arguments" that tell the program what to do. For example, -i is a switch that identifies the input file, with the file (ZOO_1080p.mov) listed next. Similarly, `-c:v` is a switch that identifies the video codec, with the codec (`libx264,` or the x264 video codec) listed next. Basically, this entire book is about FFmpeg switches you use to accomplish different things. Finally, you identify the output file (`ZOO_1.mp4`).

The command strings all assume that I'm running the batch from the same folder as the input file and that I want the output file stored in that location. Otherwise, I would insert drive: folder address information before the input and/or output file names.

Batch Files for Mac and Linux

There are two major differences between Windows and Mac and Linux batch files. First, Mac/Linux files all start with the text `#!/bin/bash.` I've seen come commentary that says the text isn't absolutely necessary, but I've always used it and it works.

```
#! /bin/bash
ffmpeg  -i ZOO_1080p.MOV -c:v libx264 ZOO_1.mp4
ffmpeg  -i TOS_1080p.MOV -c:v libx264 TOS_1.mp4
ffmpeg  -i New_1080p.MOV -c:v libx264 New_1.mp4
```

Figure 2-6. The same batch file for Mac/Linux.

The second major difference is the file extension. On the Mac, you should use the .command extension if you want to click the file to run it (as you can do with Windows). On Linux, you should use the `.sh` extension to designate the file, but that won't make the file clickable.

Introduction to Batch File Creation and Operation

Here's the basic setup and structure of running command line arguments. Note that I know just enough about batch files to automate the testing and analysis described in this and subsequent chapters. I'm certainly not a maven, and this section is designed to be an introduction to batch file operation for newbies, not an advanced course.

Plan the location of your files

I typically run batch files from the folders where the files are located. Since my primary encoding station is a 40 core HP Z840, I often run multiple encodes simultaneously. You can encode simultaneously from the same folder when you're running single-pass encodes, but it's more challenging with dual-pass encodes. That's because each first-pass creates a log file, and unless you designate the log file name in the command line, simultaneous encodes in the same folder will overwrite the log files, and hose the encodes. You can name the log files in the command line and avoid the problem, but that adds complexity. For this reason, when I'm encoding multiple files simultaneously, I encode using different folders.

Create the batch file in the native tool for each OS

Each OS uses different characters for carriage returns, so often batch files created on one OS, say Windows, won't run on another, like Linux. In addition, programs like Word often introduce funky characters that you can't see into the batch text.

So, use the simplest native application on each OS to create the batch files. On Windows, I use Notepad, on the Mac, TextEdit, and on Linus, Gedit.

Run the batch file

The technique and complexity varies by OS.

- **Windows** - Double-click the batch file or navigate to the folder in the Command line and type the batch file name.

- **Linux** - Navigate to the folder in Terminal and type . / followed by the batch file name (see Figure 2-7). Or, in Terminal, navigate to the folder containing the batch file and drag the batch file into the Terminal window. Unlike on the Mac, if Terminal isn't open to the correct folder, the batch file won't find your input file.

```
jan@Z6U:~/Desktop/CH02-Installing$ ./batch2.sh
ffmpeg version git-2017-01-22-f1214ad Copyright (c) 2000-2017 the FFmpeg develop
ers
  built with gcc 4.8 (Ubuntu 4.8.4-2ubuntu1~14.04.3)
  configuration: --extra-libs=-ldl --prefix=/opt/ffmpeg --mandir=/usr/share/man
--enable-avresample --disable-debug --enable-nonfree --enable-gpl --enable-versi
```

Figure 2-7. Running a batch file on Linux.

Note that in most, if not all cases, you'll need permission to run the batch file, and if you don't have that permission you'll see the error message Permission denied (Figure 2-8).

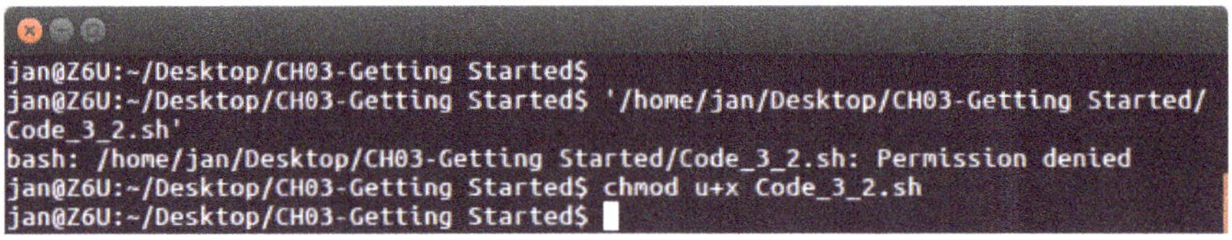

```
jan@Z6U:~/Desktop/CH03-Getting Started$
jan@Z6U:~/Desktop/CH03-Getting Started$ '/home/jan/Desktop/CH03-Getting Started/
Code_3_2.sh'
bash: /home/jan/Desktop/CH03-Getting Started/Code_3_2.sh: Permission denied
jan@Z6U:~/Desktop/CH03-Getting Started$ chmod u+x Code_3_2.sh
jan@Z6U:~/Desktop/CH03-Getting Started$ █
```

Figure 2-8. Giving permission to run the batch file.

To give the file permission to execute, type the following, which you can see in Figure 2-8.

```
chmod u+x filename
```

Then you can run the batch file again and it should execute.

- **Mac** - if the file extension is .command, you can double-click the file. Otherwise, you can navigate to the folder in terminal and type the batch file name. Or, you can drag the batch file into an open Terminal window, and it will execute the batch.

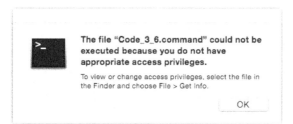

Figure 2-9. Can't run the batch file on the Mac without permission.

As with Linux, you'll often see the error message shown in Figure 2-9 when running a batch file on the Mac. To give the file permission to execute, type the following, which you can also see in Figure 2-8.

```
chmod u+x filename
```

Then rerun the batch and it should execute.

Essential Command Line Commands

When you open the Command Prompt or Terminal window, your first task is to navigate to the specific file location or location of the batch file. While there are myriad ways to do this, here are the simplest options in Windows, with links to Mac/Linux instruction to follow.

- **Change drive.** Type the drive name and colon, but not a slash. For example, type `e:` to switch to the E:\ drive, and the prompt should change to read `E:\`. If you type in the slash, you'll get an error message.

- **Change folder in a drive.** Type `cd` (for change directory), then a space (press space bar), and then the name of the directory. For example, type `cd experiment` to move to the experiment folder on the current drive. If that was the E:\ drive, you would be in E:\ experiment.

- **Navigate closer to the root directory.** To navigate closer to the root—say from E:\experiment\Test to E:\experiment—type `cd..` (cd followed by two periods). Or, from any folder, type `cd\` to move to the root of that drive (E:\).

- **View all folders in a drive.** Once you get to a drive (or folder), you may need to see what folders are in the drive. To list all folders, type `dir *.` (dir followed by *.).

- **View all files and folders in a drive.** Once you get to a drive (or folder), you may need to see the files and folders there. To do so, type `dir` (dir).

- **View all files of a specific type in a drive or folder.** Once you navigate to a drive or folder, you may need to see which files are in that folder, particularly if you're trying to run a specific batch file. Type `dir *.bat` to locate all batch files in a folder. To list all files with a different extension (like MP4 files), simply substitute in that extension (`dir *.mp4`).

- **Paste a command line argument into the command prompt.** The simplest way to test a command line argument is to copy it into the Command Prompt. Note that Ctrl + V won't work; you have to right-click in the Command Prompt window and choose Paste.

> **Tip:** *To learn how to navigate in Terminal on the Mac, check out the MacWorld article entitled, Master the command line: Navigating files and folders at bit.ly/Ter_nav_mac. To learn how to navigate in Terminal in Linux, check out bit.ly/Ter_nav_Linux. There's also a free book entitled The Linux Command Line by William Shotts that you can download at bit.ly/Ter_Linux.*

Debugging Batch Files

Unless you're exceptional, many of the batch files that you write won't run properly the first time. Sometimes it's errors in the arguments, sometimes it's something else.

When you try to run the file, the operating system may display an error message, but often it displays too quickly to comprehend. The only way to debug the file is to navigate to the batch file location in the command line or Terminal, and copy and paste the lines of arguments to the prompt and run them one by one. All this is a long way of saying if you're not skilled at navigating around and working in the command line or Terminal, you're going to have a frustrating time running FFmpeg.

OK, now that we've got FFmpeg installed, let's start encoding some files!

Chapter 3: Choosing Codecs and Container Formats

```
ffmpeg  -i TOS_1080p.mov -c:v libx264 TOS_1080p_win.mp4
ffmpeg  -i TOS_1080p.mov -s 1280x720 -c:v libx264 TOS_720p_win.mp4
ffmpeg  -i TOS_1080p.mov -s 640x360 -c:v libx264 TOS_360p_win.mp4
```

Batch 3-1. Choosing the H.264 video codec in FFmpeg.

In this chapter, you'll learn how to choose a codec and container format for your FFmpeg encoded videos. Even more importantly, you'll learn which parameters FFmpeg assigns when you don't specify a specific parameter in the command string, which will become increasingly important as we move through the book. Specifically, here's what you will learn in this chapter.

- how to choose a codec with FFmpeg

- which parameters FFmpeg chooses when you don't specifically assign them in the default string

- how to choose a container format in FFmpeg

Designating the Codecs in FFmpeg

Here are the components of the command strings shown in Batch 3-1.

ffmpeg.exe calls the program.

-i TOS_1080p.mov name of input file.

-s 1280x720 changes the resolution of the output file.

-vcodec libx264 chooses the video codec, in this case the x264 codec.

TOS_1080p.mp4 output file name.

The string calls FFmpeg, and tells it to convert TOS_1080p.mov using the x264 codec and storing it as TOS_1080p.mp4. I changed the resolution in the second two files to see if FFmpeg applied different defaults based upon resolution. You'll learn multiple options for changing the resolution in Chapter 5.

Table 1 shows the specs of the input file and the configuration of the three files produced with Batch 3-1, which is about the most basic FFmpeg command string possible. If you're new to streaming video, a lot of these parameters will be meaningless to you. As you'll learn a bit below, and in more detail throughout the book, while many of these parameters are fine for single file streaming, they don't work for files produced for adaptive bitrate (ABR) streaming, which is the focus of much FFmpeg encoding today.

	Input	1080p Output	720p Output	360p Output
Container	MOV	MP4	MP4	MP4
Codec	H.264	H.264	H.265	H.266
Resolution	1920x1080	1920x1080	1280x720	640x360
Frame Rate	24	24	24	24
Profile	High	High	High	High
CABAC	Yes	Yes	Yes	Yes
Data Rate	30 mbps	3569	1683	832
Max Rate	30 mbps	4898	2436	1431
Keyframe Interval	3 seconds	250 frames	250 frames	250 frames
Keyframes at Scene Changes	Yes	Yes	Yes	Yes
Preset	Medium	Medium	Medium	Medium
B-frame	2	3	3	3
R-frame	3	3	3	3
Audio Bitrate	317 kbps	132 kbps	132 kbps	132 kbps
Channels	Stereo	Stereo	Stereo	Stereo
Sample Rate	48 kHz	48 kHz	48 kHz	48 kHz

Table 3-1. The output from minimal FFmpeg command string.

Let's start with some observations about how FFmpeg encoded the files. Since we didn't change the resolution of the 1080p output file, so let's focus our attention there.

- **Resolution and frame rate.** FFmpeg doesn't change the resolution or frame rate of the input file unless directed to do so. If you're building an encoding ladder like the one shown in Table 1-2, you'll need to manually change the resolution in your command strings. You'll learn about resolution in Chapter 5, and frame rate in Chapter 6.

- **Profile and entropy coding.** FFmpeg seems to default to the High profile in all cases unless otherwise directed to do so, and to enable CABAC entropy coding. If you need files encoded using the Baseline or Main profile, you'll have to do that manually. You'll learn about these H.264-specific parameters in Chapter 9.

- **Data rate and data rate control.** As you learned in Chapter 1, data rate largely controls quality and cost, since you may be paying for the bandwidth that you consume. In most cases, you'll want to set the data rate manually, which you'll learn how to do in Chapter 4.

- **Keyframe interval and scene change detection.** Unless directed otherwise, FFmpeg defaults to a keyframe interval of 250 frames and inserts keyframes at scene changes.

Files encoded for ABR distribution must have regular keyframes at much shorter intervals that match the duration of segments used by all ABR techniques. You'll learn how to accomplish this in Chapter 7.

* **Preset.** The encoding preset largely controls the tradeoff between quality and encoding time. By default, FFmpeg uses the Medium preset, which generally strikes a good balance. However, you can shorten encoding time with minimal impact on encoding quality by using a different preset or gain a bit of quality by using a different preset that extends encoding time. You'll learn all about this in Chapter 8.

* **B-frame and reference frames.** These parameters are controlled by the default Medium preset deployed by FFmpeg. There are some instances where you may wish to customize these, which you'll learn about in Chapter 7.

* **Audio parameters.** Unless otherwise directed, FFmpeg appears to reduce the audio bitrate to around 128 kbps, while keeping channels (stereo) and sample rate (48 kHz) the same. Some producers like to customize audio parameters for lower bitrate encodes which you'll learn how to do in Chapter 9.

Interestingly, other than resolution and data rate, FFmpeg used the same parameters for the 720p and 360p files as the 1080p file. I thought that FFmpeg used the Baseline or Main profiles by default for lower resolution files, but this turned out not to be the case, at least in these limited tests.

The essential point is this. Simple command strings can produce files that are perfectly suitable for casual playback from your hard disk, but you'll have to get much more specific to produce files for streaming, and particularly adaptive bitrate delivery.

Other Codecs

To choose a video codec in FFmpeg, you use the `-c:v` or `-vcodec` string, followed by the identity of the codec. Here are the video codecs you'll learn work with in this book.

`-c:v libx264` H.264 using the x264 video codec. You'll learn about H.264-specific parameters in Chapter 10.

`-c:v libx265` HEVC using the x265 video codec. You'll learn about this in Chapter 13.

`-c:v libvpx-vp9` VP9 video codec, which you'll learn about in Chapter 13.

To designate an audio codec in FFmpeg, you use the `-c:a` or `-acodec` command. Here are the audio codecs you'll work with in this book.

`-c:a aac` the AAC audio codec for H.264 and HEVC

`-c:a libopus` the Opus codec for VP9.

Designating the Container Format in FFmpeg

To select a container format in FFmpeg, use the `-f` command. Note that you don't have to include this command in single line arguments, as FFmpeg will infer the container format from the file extension of the output file. That's why it's not included in Batch 3-1, yet this produces a perfectly fine MP4 file in the MP4 container format.

With two-pass encoding, there is no output file name in the first line, so there's no file extension to infer the container from. So, you must designate the container format in the first line of two-pass encodes. Here are the container formats you'll work with in this book.

`-f MP4` MPEG 4 container format for H.264 and HEVC

`-f webm` WebM container format for VP9

`-f hls` HTTP Live Streaming container format

Changing the Container Format in FFmpeg

FFmpeg is an awesome tool for changing the container format of file. For example, suppose you had a file in QuickTime (.mov) format that you needed in MP4 format (.mp4). Assuming the file was encoded with H.264/ACC, the command would be:

```
ffmpeg -i mov.mov -vcodec copy -acodec copy mp4.mp4
```

Batch 3-2. Changing from QuickTime to MP4 container.

Here, FFmpeg copied the compressed audio/video streams into a new file using the MP4 container format. You can see this by comparing the two MediaInfo boxes shown together in Figure 3-1, MOV on the left, MP4 on the right. The only differences relate to the container format shown in the top section; the video parameters are identical, as are the audio (not shown). This is obviously a lossless conversion since there is no re-encoding.

Tip: The tool shown in Figure 3-1 is called MediaInfo and it's indispensable for any video producer. It's available for free download at bit.ly/DL_MI and runs on Windows, Mac, and Linux. For a tutorial on MediaInfo and a tool you will meet in Chapter 4, Bitrate Viewer, check out bit.ly/videoanalyze.

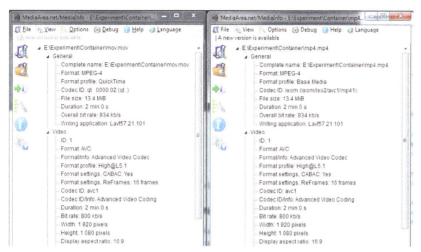

Figure 3-1. Changing the container format from MOV to MP4.

You can also convert MPEG-2 transport streams (.ts) into MP4 files with FFmpeg, but you must include the `-bsf: a aac_adtstoasc` string shown below.

```
ffmpeg -i MPEG.ts -bsf:a aac_adtstoasc -c:v copy -c:a copy mp4.mp4
```

Batch 3-3. Changing from an MPEG-2 transport stream to an MP4 file.

This particular conversion came in ultra-handy in a recent consulting project that involved dozens of files encoded into the MPEG-2 transport stream container format. Re-encoding would have taken hours, if not days; the simple change in container format (also called transmuxing) took minutes.

OK, now we know how FFmpeg sets its default values, so now let's learn how to change them to what we need.

Chapter 4: Bitrate Control

Figure 4-1. Sometimes CBR-encoded video exhibits transient quality glitches like this one in the movie Zoolander.

Whenever you encode a file, you must choose both the bitrate and the bitrate control technique, or how the video data rate is allocated within the file. For most producers, this means a choice between constant bitrate (CBR) or variable bitrate (VBR) encoding. While these choices have been available since the dawn of H.264 (and MPEG-2 and before it, for that matter) there still is no consensus as to which is best to use.

In general, the CBR-versus-VBR decision involves a debate between quality and deliverability. It's generally accepted that VBR produces better quality than CBR, although it probably doesn't make as big a difference as you might think. Despite the quality advantage, many producers use CBR over concerns that variances in the VBR bitrate will make their files harder to deliver, particularly over constrained bitrate connections like 3G and 4G. As you'll learn later in this chapter, these concerns are appropriate. Otherwise, in this chapter, you will learn:

- how VBR and CBR work

- differences in overall frame quality

- how both techniques affect deliverability

- what the Video Buffering Verifier (VBV) is and how it affects bitrate control and quality

- best practices for encoding with CBR and VBR

- how to encode CBR and VBR files in FFmpeg

- what CRF encoding is and how to use it.

CBR and VBR Defined

Most encoding tools provide the bitrate control options shown in Figure 4-2, CBR or VBR. As you'll learn in this chapter, you can encode using both techniques with FFmpeg as well.

Figure 4-2. What will it be today, CBR or VBR?

Let's use the file shown in Figure 4-3 to illustrate the difference between CBR and VBR. As you can see, the file has fives scenes, as follows:

- **Low motion.** Talking head

- **Moderate motion.** Woman cooking pita bread on an outdoor oven

- **Low motion.** An integrated-circuit chip-cutting machine in operation

- **Moderate motion.** A musician playing the violin

- **High motion.** Walking holding the camcorder to my chest and panning side to side.

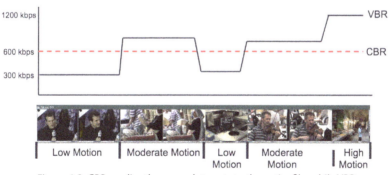

Figure 4-3. CBR applies the same data rate to the entire file, while VBR varies the data rate to match scene complexity.

With CBR, the dotted red line, the bitrate is constant throughout, a flat line at 600 kbps. In contrast, VBR varies the bitrate according to the complexity of the video—lower in the easy-to-compress talking-head sequence, and higher in the high-motion sequence at the end.

Note that you can produce CBR using either a single pass or multiple passes. With multiple passes, the encoder assesses complexity during the first pass, and then encodes and allocates bits during the second. Obviously, with live encodes (not addressed in this book), CBR is produced in a single pass.

Almost all VBR is produced using two passes, again, one for analysis, one for encoding. In addition, most VBR is "constrained," which means you assign a maximum data rate that the encoder won't exceed. So, 200% constrained VBR means a maximum data rate of 200% of the target, while 110% constrained VBR means a maximum rate of 110% of the target. You should constrain all VBR encodes produced for streaming because if data rate spikes get too high, you may experience problems delivering the files under constrained conditions like 3G and 4G.

When choosing between VBR and CBR, you should consider three elements, overall quality, transient quality, and file deliverability.

1. Overall quality

First is overall quality, which is shown in Table 4-1. As with most tables in this book, the number with the green background is the best quality, while the number with the red background is the worst. As you can see, 200% constrained VBR delivers the best quality in almost all cases, while one or two pass CBR delivers the worst in seven of eight files.

PSNR	200% VBR	150% VBR	110% VBR	CBR 2-Pass	CBR 1-Pass	Total Quality Delta	Delta - 110% to 200%
Tears of Steel	41.97	41.89	41.60	41.40	41.41	1.36%	0.88%
Sintel	41.34	41.13	40.67	40.56	40.17	2.83%	1.64%
Big Buck Bunny	41.73	40.98	40.00	39.70	40.07	4.88%	4.14%
Talking Head	44.23	44.22	44.17	44.12	44.15	0.25%	0.14%
Freedom	42.06	42.02	41.84	41.83	41.65	0.98%	0.53%
Haunted	42.07	42.07	42.01	41.90	42.06	0.40%	0.15%
Tutorial	46.81	46.56	45.27	45.08	44.71	4.49%	3.29%
Screencam	39.71	38.31	36.89	36.96	40.01	7.80%	7.11%
1080p Average	42.49	42.15	41.56	41.44	41.78	2.46%	2.20%
720p Average	41.35	41.18	40.82	40.77	40.76	1.71%	1.28%

Table 4-1. PSNR quality using different bitrate control techniques.

On the other hand, the quality difference in the Total Quality Delta column averages only 2.46% for 1080p video, and 1.71% for 720p video, which few, if any, viewers would notice. So, while VBR is widely (and accurately) touted as delivering the best possible quality, the difference isn't as dramatic as you might think.

2.   Transient quality.

The big issue with CBR is that quality can drop precipitously for one or two frames in high-motion sequences (Figure 4-1). While this doesn't happen all that frequently, it's still the most important reason to avoid CBR.

3.   Deliverability

The final consideration for choosing a bitrate control technique is the ability to deliver the file over constrained connections. This is shown in Figure 4-4, which shows two views of an application called Bitrate Viewer. The top view shows the bitrate of a CBR-encoded file, which is relatively flat, the bottom VBR-encoded file, which shows significant variances in data rate. Which would you rather deliver over a 3G connection?

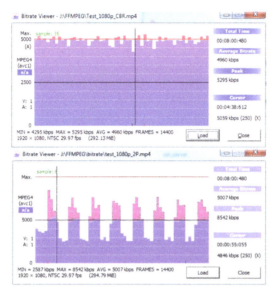

Figure 4-4. CBR file on top, VBR on the bottom.

I examined the impact of data rate control technique in an article on the *Streaming Learning Center* entitled, Bitrate Control and QoE-CBR is Better, which you can read at bit.ly/vbr_cbr_qoe. Using a specially constructed file that contained 30 seconds of talking head followed by 30 seconds of high-motion ballet footage, I showed how 200% constrained VBR can seriously degrade the quality of experience (QoE) of viewers on constrained connections.

Overall, the article recommends that you produce using between 110% and 150% constrained VBR for encodes produced for streaming. Note that there are some contrary views. For example, Apple recently changed their recommendation for HTTP Live Streaming from 110% constrained VBR to 200% constrained VBR in TN2224, though they cited no QoE-related data.

On the other side of the coin, there are still many producers who swear by CBR. Overall, given the lack of significant quality differences between 110% and 200% constrained VBR, and the deliverability risk proven by the aforementioned article, I think 110%-150% is the safer choice.

There's one more concept you need to understand before tackling data rate control using FFmpeg. That's VBV, and it's our next stop.

A Quick Word on VBV

VBV stands for Video Buffering Verifier, and it refers to how much video data is stored (or cached) in the player. As you'll see, when setting the bitrate with FFmpeg, you'll set the target bitrate, maximum bitrate, and VBV buffer size.

In general, the larger the buffer size, the higher the quality and the greater the variability in data rate within the stream. If you're encoding with CBR, and need a CBR stream, you must keep the buffer small, usually the same size as a single second of video data, which will become crystal clear in a moment. If you're producing using 200% constrained VBR and don't necessarily care about data rate consistency, using 2 seconds of data is acceptable.

Bitrate Control and Buffer Size in FFmpeg

Implementing the bitrate control technique and buffer size in FFmpeg is simple. To illustrate how, and the effect of each technique, I'll use the test video file that I created for the QoE article mentioned above, which again, was eight minutes long, and alternates 30 seconds of talking head video with 30 seconds of high-motion ballet.

To set the bitrate target in FFmpeg, use the `-b:v` code (bitrate:video) below :

```
ffmpeg -i Test_1080p.MP4 -c:v libx264 -b:v 5000k Test_DR_5M.mp4
```

Batch 4-1. Producing a file using the `-b` *command and one-pass encoding.*

This produces a file that looks like Figure 4-5, where the data rate would vary according to content. The overall data rate of 5062 is pretty accurate, but you'd be concerned that data rate spikes in the file could hinder deliverability. The answer? Two-pass encoding.

How do you control data rate with two-pass encoding? Using two new controls, maximum bitrate and VBV buffer size. That is, you use:

`-b:v 5000k` as the target, as before.

`-maxrate 5000k` to set the maximum bitrate. So, 5000k would be CBR, 5500k would be 110 percent constrained VBR, and 10000k would be 200 percent constrained VBR.

`-bufsize 5000k` to set the size of the VBV. For this real-world video distributed via streaming, I'd use a VBV size equivalent to the data rate of one second of video (5000k).

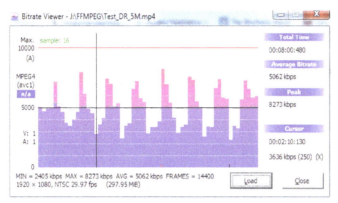

Figure 4-5. One-pass encoding at 5 Mbps.

Tip: *Note that if you forget the "k" in the bitrate (5000***k***), FFmpeg will encode to bytes, not kilobytes. If you find your encoded files abnormally small, it's likely that you forgot the k.*

Two-Pass Encoding in FFmpeg

To implement two-pass encoding in FFmpeg, you define both passes, each in their own line. This is what the two lines would look like.

```
ffmpeg -y -i  test_1080p.mp4 -c:v libx264  -b:v 5000k -pass 1  -f mp4
NUL && \

ffmpeg -i  test_1080p.mp4 -c:v libx264  -b:v 5000k -maxrate 5000k
-bufsize 5000k -pass 2  test_1080p_CBR.mp4
```

Batch 4-2. Producing a CBR file with two-pass encoding.

Here is a description of the new controls added to the command line.

Line 1. During this pass, FFmpeg scans the file and records complexity in a log file.

`-y` overwrites existing log file. If you encode multiple files using FFmpeg, this tells the program to overwrite the existing log file. Without `-y`, FFmpeg will stop the batch to ask if you want to overwrite the log file each encode. Or you can name the log file for each encode with the `-passlogfile` switch.

`-pass 1` completes the first pass and creates the log file but no output file.

`-f mp4` identifies the output format used in the second pass.

`NUL` creates the log file.

`&& \` tells FFmpeg to run a second pass if the first pass was successful.

Line 2. During this pass, FFmpeg uses the log created in the first pass to encode the file.

`-b:v 5000k` sets the overall target.

`-maxrate 5000k` sets the maximum bitrate. It's the same as the target, so this means CBR.

`-bufsize 5000k` sets the size of the VBV.

`-pass 2` finds and uses the log file for the encode.

`test_1080p_CBR.mp4` sets the output file name.

Note that all bitrate-related commands in the first pass must also be in the second pass; you'll learn more about that in Chapter 10. Figure 4-6 shows the CBR-encoded file in Bitrate Viewer. Although the file isn't a total flat line, there's much less data rate variability than in Figure 4-5, and the file would be much simpler to deliver. Of course, overall quality is slightly lower than VBR, and there's a risk of transient quality problems.

Figure 4-6. Two-pass CBR encoding with FFmpeg.

Quick Summary: Constant Bitrate Encoding

1. CBR delivers the lowest quality stream with occasional transient issues but is the easiest stream to deliver.

2. You produce a CBR stream by using the same value for target and maximum bitrate in either a single or two-pass encode.

3. When producing CBR files, you should use a VBV buffer of one-second of video.

200 Percent Constrained VBR Encoding in FFmpeg

Here's how to produce 200% constrained VBR in FFmpeg. The first line is the same, but I've boosted -maxrate to 10000k in the second line.

```
ffmpeg -y -i  test_1080p.mp4 -c:v libx264  -b:v 5000k -pass 1  -f mp4
NUL && \

ffmpeg -i  test_1080p.mp4 -c:v libx264  -b:v 5000k -maxrate 10000k
-bufsize 5000k -pass 2  test_1080p_200p_CVBR.mp4
```

Batch 4-3. Producing a 200% constrained VBR file with two-pass encoding.

Figure 4-7 shows the 200 percent constrained VBR file in Bitrate Viewer. Quality would be optimal, and there should be no transient quality problems. Again, however, with this worst-case file with mixed high- and low-motion footage, deliverability might be a real issue.

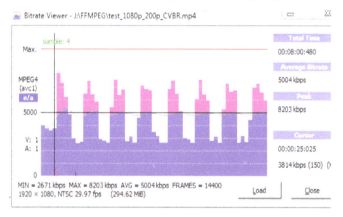

Figure 4-7. Two-pass 200 percent constrained VBR encoding with FFmpeg.

110 Percent Constrained VBR Encoding in FFmpeg

Here's how to produce 110% constrained VBR in FFmpeg. The first line is the same as the previous two, but -maxrate in pass two is limited to 5500k.

```
ffmpeg -y -i  test_1080p.mp4 -c:v libx264  -b:v 5000k -pass 1  -f mp4
NUL && \

ffmpeg -i  test_1080p.mp4 -c:v libx264  -b:v 5000k -maxrate 5500k
-bufsize 5000k -pass 2  test_1080p_110p_CVBR.mp4
```

Batch 4-4. Producing 110% constrained VBR file with two-pass encoding.

Figure 4-8 shows the 110% constrained VBR file in Bitrate Viewer. The data rate is very similar to the other two, of course—although the peak bitrate is 5852 kbps compared with 5295 for

CBR. While quality would be slightly less than 200 percent constrained VBR, there should be no transient quality problems, and the file should be pretty simple to deliver.

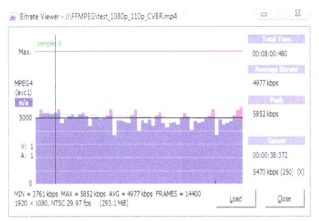

Figure 4-8. Two-pass 110 percent constrained VBR encoding with FFmpeg.

Tip: *Running multiple FFmpeg encodes simultaneously is a great way to speed up your multiple file encoding chores—particularly on a multiple-core computer. Be careful when encoding multiple files in the same folder using two-pass encoding, however, since you'll be creating multiple log files that will overwrite each other and ruin the second encode. You can separately name the log file using the* -passlogfile *switch, or simply run the different encodes from different folders, which is what I do.*

Quick Summary: Variable Bitrate Encoding

1. VBR delivers the highest quality stream with very few transient issues, but data rate swings can complicate delivery and degrade QoE.

2. All VBR encodes should be constrained by limiting the maximum bitrate. I recommend a maximum setting of 110-150% of the target.

3. All VBR encodes should be two-pass.

4. When producing VBR files for streaming, use a VBV buffer of between one and two seconds of video.

Constant Rate Factor (CRF) Encoding

When you encode using CBR and VBR, you choose a data rate and bitrate control technique, and FFmpeg attempts to meet that data rate using the selected bitrate control technique. With Constant Rate Factor (CRF) encoding, you choose a quality level and FFmpeg delivers that quality, adjusting the data rate up and down as needed. You get a file with a fixed quality level, but unknown (in advance) data rate, and a file where the data rate varies significantly over the duration of the file, which may impact deliverability.

Using CRF

Figure 4-9 shows how CRF values affect quality; specifically, the lower the value, the higher the quality. It's counter-intuitive, but that's how it works.

Figure 4-9. How CRF values affect video quality. From the CRF Guide.

How is CRF useful? Two ways. First, it's a measure of file complexity. That is, if you encode a talking head clip and a soccer clip using the same CRF value, the soccer clip will have a much higher data rate. That's because the higher level of motion requires more data to achieve the same quality level.

Some encoding systems use CRF values as a gauge to compute the data rate necessary to produce sufficient quality for each clip. With the talking head and soccer clips, these systems would run a CRF encode, measure the data rate, and then encode using CBR or VBR at the calculated data rate to achieve quality and maintain deliverability.

The second way CRF is useful is as a bitrate control technique with a "capped" data rate, which is called capped CRF. It's almost easier to show than explain, so let's jump in with the FFmpeg controls.

With CRF encoding, you insert a CRF value rather than a data rate as shown in Batch 4-5. Since the goal is quality, not a data rate target, all CRF (and capped CRF) encodes are single pass.

```
ffmpeg -i Test_1080p.MP4 -c:v libx264 -crf 23 Test_CRF23.mp4
```

Batch 4-5. Producing a file with a CRF value of 23.

-crf 23 tells FFmpeg to encode using a CRF value of 23. Note that in *Encoding by the Numbers*, we learned that a CRF value of 23 approximates the data rate (and quality) delivered by most Hollywood producers. That's why I used this value here.

As you can see, you swap the -crf command for the data rate controls, which produces the file shown in Figure 4-10. As you can see, the data rate varies from around 2400 for the talking head sections to around 4500 kbps for the ballet sections. The average data rate is 3153, about 40% less than the 5 Mbps used in previous encodes.

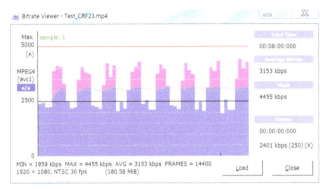

Figure 4-10. Test file encoded at CRF 23.

This is the attraction of CRF encoding; it applies the data rate necessary to preserve quality, and that's it. The problem is, of course, we need a maximum data rate to ensure file deliverability.

Capped CRF

If we capped the data rate at 5 Mbps, the file would look very similar to Figure 4-10 because there's no section where the data rate would be capped–it never exceeds 5 Mbps. So, let's cap it at 3500k with the following command string.

```
ffmpeg -i Test_1080p.mp4 -c:v libx264 -crf 23 -maxrate 3500k -bufsize 3500k
Test_CRF23_3500.mp4
```

Batch 4-6. Producing a file with a CRF value of 23 and a cap of 3500 kbps.

`-crf 23` sets the CRF level.

`-maxrate 3500k` sets the maximum data rate.

`-bufsize 3500k` sets the buffer size.

Figure 4-11. Test file encoded at CRF 23 and capped at 3500 kbps.

If you compare Figure 4-10 and 4-11, you'll see that the talking head sections are about the same data rate, but the maximum data rate has been restricted to around 3500, as requested, and the overall data rate reduced from 3153 kbps to 2758 kbps.

Note that some companies use capped CRF for distribution, including online video platform vendor JWPlayer, who uses it for both H.264 and VP9. The downside is the irregularity of the data rate and the potential degradation of file deliverability. But JWPlayer is very credible and I'm sure they wouldn't use capped CRF if it degraded file deliverability for their customers.

OK, that's it for data rate control, next up is setting resolution.

Chapter 5: Setting Resolution

If you don't set video resolution in the command string, FFmpeg will output the same resolution as the input file. Sometimes, you want this, sometimes you don't. There are multiple ways to set video resolution in FFmpeg, and I'll cover five of them. Specifically, in this chapter, you will learn:

- how to directly set video resolution via the -s command

- the difference between pixel aspect ratio (PAR) and display aspect ratio (DAR) and why you care

- how to set target width and have FFmpeg compute the height (and vice versa)

- how to control cropping and letterboxing in FFmpeg

Setting Resolution in FFmpeg

As you learned back in Chapter 1, video resolution is the width and height of the video file in pixels. So, 1280x720 resolution means a file that's 1280 pixels wide, and 720 high. In most cases, we start production with a 1080p or 4K file which we scale down to smaller resolutions for various rungs on our encoding ladder. The simplest technique for this is the -s switch.

-s for Simple

This technique uses the -s switch to scale the video to the target resolution. As an example, Batch 5-1 scales the 1920x1080 source file to 1280x720 resolution.

```
ffmpeg.exe -i TOS_1080p.mov -s 1280x720 TOS_720p_out.mp4
```

Batch 5-1. Scaling to 1280x720 using the most simple technique.

Since the input and output files both have the same 16:9 aspect ratio, FFmpeg can scale the output file without changing the aspect ratio, which is the simplest case. How can you tell they're both 16:9? Because 1920÷16=120, and 120x9 = 1080, and 1280÷16=80 and 80*9=720.

The -s switch works best when the input and output files share the same aspect ratio. When they don't, you risk distorting the pixels. Understanding why and learning how to avoid these problems involves diving into the rabbit hole of display and pixel aspect ratios.

Pixel Aspect Ratio and Display Aspect Ratio

Suppose we had a file that didn't have an aspect ratio of 16:9, but we wanted to output a file with a 1280x720-ish resolution. For example, Blender movie Tears of Steel was produced at 3840x1714 resolution, as shown in Figure 5-1.

Figure 5-1. Tears of Steel has a resolution of 3840x1714, and a 2.25:1 display aspect ratio.

The answer surprisingly complex and requires an understanding of the difference between the display aspect ratio and pixel aspect ratio. Figure 5-2, another shot of MediaInfo, will help us understand the difference.

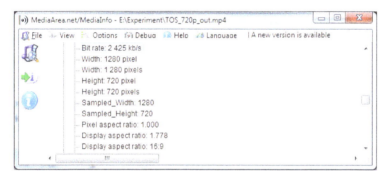

Figure 5-2. MediaInfo in Advanced Mode (Debug > Advanced Mode) showing aspect ratio data.

This is the 1280x720p file produced with Batch 5-1 which encoded a 1080p version of the Tears of Steel video file. You can see in Figure 5-2 that the resolution of the output file is 1280x720, which is what we selected via the `-s 1280x720` command. You also see values for the pixel aspect ratio (1.0) and display aspect ratio of 16:9.

Briefly, the *pixel* aspect ratio *(PAR)* tells the player how to display the pixels. A PAR of 1.0 tells the player to display each pixel one pixel wide, and one pixel high, which is also called square pixels. The *display* aspect ratio *(DAR)* is the aspect ratio of the actual display, or 16:9.

When producing for streaming, your goal is to create a file with a square pixel ratio of 1.0; otherwise the player may scale the file during display. Again, as mentioned above, the `-s` command only delivers this if the source and output file share the same PAR. If they don't, you may have a problem.

```
ffmpeg.exe -i TOS_4K.mp4 -s 1280x720 TOS_720p_S.mp4
```

Batch 5-2. Scaling to a different display aspect ratio using the simplest technique.

For example, let's see what happens when we use the `-s` switch to produce a 16:9 aspect ratio file from the non-16:9 aspect ratio Tears of Steel 4K source. That would be Batch 5-2, which produces a file with the characteristics shown in Figure 5-3 and Figure 5-4. Note that this is a different input file than the one used in Batch 5-1, which was a 1080p version of TOS with a 16:9 aspect ratio. Batch 5-2 encodes the original 4K file, which has a display aspect ratio of 2:25:1.

Figure 5-3. PAR is set to 1.260, so the video player will stretch the video by that amount.

As you can see, even though the resolution is correct in Figure 5-3 (1280x720), the PAR is set to 1.260, which tells the player to stretch each pixel to 1.26 pixels wide during display. If we load the file in QuickTime Player, you'll see what I mean (Figure 5-4). In the Movie Inspector box, you see the Normal Size is 1613x720. How do you calculate 1613? Multiply 1280 times the 1.26 aspect ratio and you get 1613.8. So, QuickTime stretched the file by 1.26x

Figure 5-4. QuickTime scaled the 1280 width by 1.26.

In essence, FFmpeg is preventing you from distorting the file by maintaining the display aspect ratio of the source. If you do the math (1613 divided by the display aspect ratio of 2.24 shown in

Figure 5-4), you get 720, the height of the file. So 1613x720 has a display aspect ratio of 2.24:1, which is almost exactly the same aspect ratio as the original 3840x1714 file (2.25:1).

In practical terms, using `-s`, in this case, will either tell the player to scale the file outside the bounds of the player, or will distort the pixels to play the file in the 1280x720 window. To avoid this, we must use a command that delivers the desired resolution and a PAR of 1.0, or very close thereto. The `-s` command doesn't get this done. What's the best alternative? It's one of four choices, depending upon your goal.

- Target width (1280) and compute height

- Target height (720) and compute width

- Target 1280x720 resolution and crop excess pixels

- Target 1280x720 resolution, letterbox and retain all pixels

Let's look at each in turn.

> **Note:** *For a background article (more than you ever wanted to know) about square pixels, check out* bit.ly/square_pix.

Target Width and Compute Height (1280 x ?)

Use this alternative to produce files with a 1280 width, a computed height that preserves the original aspect ratio, here 2.25:1, and a PAR of 1.0.

```
ffmpeg -i TOS_4K.mp4 -c:v libx264 -vf scale=1280:trunc(ow/a/2)*2
TOS_VF_trunc_1280_proof.mp4
```

Batch 5-3. Scaling to 1280 and computing height.

Here's an explanation of the new commands in this string.

> `-vf` calling a video filter in FFmpeg. Technically, a video filter is a tool that modifies the input before it's encoded. Here, we're using a filter to scale the video.
>
> `scale=1280:trunc(ow/a/2)*2` setting the width at 1280, and computing the height by dividing the output width by the aspect ratio, then dividing by two and multiplying by two to ensure at least mod-2 output. To use this approach, substitute your target width for 1280.

Figure 5-5 tells us if we met our goals. On the left, we see that the file has an output width of 1280 and a height of 570. Do the math (1280÷570) and you get 2.241, just a bit off from the original aspect ratio of 2.25. The pixel aspect ratio is .998, which very close to the target of 1.0. On the right, we see that this file plays in QuickTime as 1277x570, showing that FFmpeg produced very close to the desired result.

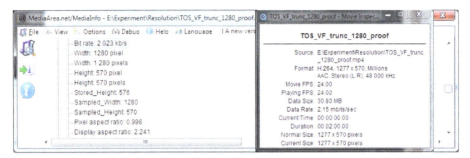

Figure 5-5. MediaInfo and QuickTime info.

Target Height Compute Horizontal (? x 720)

Use this alternative to produce files with a 720 height, a computed width that preserves the original aspect ratio, here 2.25:1, and a PAR of 1.0.

```
ffmpeg -i TOS_4K.mp4 -c:v libx264 -vf scale=trunc(oh*a/2)*2:720
TOS_VF_trunc_720_proof.mp4
```

Batch 5-4. Scaling to 720 height and computing width.

Here's an explanation of the new commands in this string.

> `-vf` calling a video filter in FFmpeg.

> `scale=trunc(oh*a/2)*2:720` setting the height at 720, and computing the width by multiplying the output height times the aspect ratio, then dividing by two and multiplying by two to ensure at least mod-2 output. To use this approach, substitute your target height for 720.

Figure 5-6 tells us if we met our goals. On the left, we see that the file has an output height of 720 and a width of 1612. Do the math (1613/720) and you get 2.241, just a bit off from the original aspect ratio of 2.25. The pixel aspect ratio is 1.001, which very close to the target of 1.0. On the right, we see that this file plays in QuickTime as 1613x720, so again, mission accomplished.

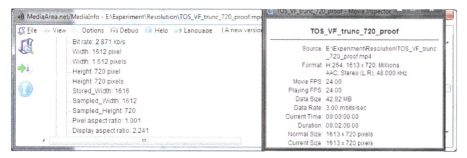

Figure 5-6. MediaInfo and QuickTime info.

Target 1280x720 resolution and Crop Excess Pixels

Use this alternative to produce files with a 1280x720 resolution, with excess pixels cropped. Output display aspect ratio will be 16:9, and pixel aspect ratio will be 1.0.

```
ffmpeg -i TOS_4K.mp4 -vf
"scale=1280:720:force_original_aspect_ratio=increase,crop=1280:720"
TOS_1280_crop.mp4
```

Batch 5-5. Scaling to 1280x720 and cropping excess pixels.

Here we're using a video filter again, but with two commands, scale and crop. So, use quotation marks at the start and end to let FFmpeg know that there are two commands. The only numbers you would have to change to use this approach are the width and height.

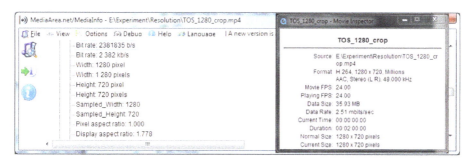

Figure 5-7. MediaInfo and QuickTime info.

Figure 5-7 shows that we met our goals. On the left, we see that the file has a resolution of 1280x720, with a display aspect ratio of 1.778 (which is 16 divided by 9), and a pixel aspect ratio of 1.000. On the right, we see that this file plays in QuickTime as 1280x720, which is just what we want.

Figure 5-8 shows what was cropped from the video. While not insubstantial, this is probably the approach most producers take when working with 4K videos that don't have a 16:9 display aspect ratio.

Figure 5-8. This shows what was cropped using the previous command string.

Target 1280x720 resolution and Letterbox

Use this alternative to produce files with a 1280x720 resolution, with all pixels squeezed into the video via a letterbox on top and bottom of the video. The output display aspect ratio will be 16:9, and pixel aspect ratio will be 1.0.

```
ffmpeg -i TOS_4K.mp4 -vf
"scale=1280:720:force_original_aspect_ratio=decrease,pad=1280:720:(ow-
iw)/2:(oh-ih)/2" TOS_1280_padding.mp4
```

Batch 5-6. Scaling to 1280x720 and letterboxing.

Here we're using a video filter again, also with two commands, scale and pad, so we use quotation marks at the start and end. The only numbers you must change to use this approach are the width and height.

Figure 5-9. MediaInfo and QuickTime info.

Figure 5-9 shows that we met our goals. On the left, we see that the file has a resolution of 1280x720, with a display aspect ratio of 1.776 (which is 16 divided by 9), and a pixel aspect ratio of .999. On the right, we see that this file plays in QuickTime as 1279x720, so yet again, FFmpeg achieved the desired result.

Figure 5-10 shows the letterboxed video we just created. If you want to keep all the action in the frame, and still fit the video into a 1280x720 window, this is the approach that you should take.

Figure 5-10. This shows the letterboxed video fit into a 1280x720 window.

Note: *I based the crop and letter box approaches shown above on an article entitled, Resizing videos with ffmpeg/avconv to fit into a static sized player, on SuperUser. For more details about this approach, you can find this article at* bit.ly/crop_lbox.

So that's video resolution. In the next chapter, you'll learn how to choose the appropriate frame rate for a file, and how to set that frame rate in FFmpeg.

Chapter 6: Setting Frame Rate

Video average bit rate (kb/s)	Resolution	Frame rate
145	416 x 234	≤ 30 fps
365	480 x 270	≤ 30 fps
730	640 x 360	≤ 30 fps
1100	768 x 432	≤ 30 fps
2000	960 x 540	same as source
3000	1280 x 720	same as source
4500	same as source	same as source
6000	same as source	same as source
7800	same as source	same as source

Table 6-1. Consider reducing the frame rate in the lower rungs of your encoding ladder.

If you don't set the frame rate in your command string, FFmpeg will output the same frame rate as the input file. In most instances, this is the correct decision, but sometimes not. In this chapter, you will learn:

- when to consider changing the frame rate of your encoded videos

- how to do so in FFmpeg.

Overview

In Chapter 1, you learned that data rate, resolution, and frame rate all impact the quality of your encoded file. In most video configurations, you use the same frame rate as the source, and control quality using resolution and data rate. When should you consider reducing the frame rate? Only for the lowest rungs on your encoding ladder.

This is shown in Table 6-1, which is from Apple's HLS Authoring Specification (bit.ly/A_Devices_Spec). While Apple directs you to use the same frame rate as the source in higher quality files, you can use 30 fps or below for the bottom four files. In my experience, few producers reduce the frame rate of files configured at 640x360 or higher, leaving the bottom two as the prime candidates.

When to Cut the Frame Rate?

Theoretically, halving the frame rate doubles the data applied to each pixel in the file. When working at the lowest rungs in the encoding ladder, this can improve quality considerably.

Cutting the frame rate may also allow you to use a higher resolution than you otherwise would at 30 fps, eliminating the blockiness and blurriness that occurs when you scale low-resolution video to fit the player window, or worse yet, full screen. You see this in Figure 6-1, where the left image is from a higher resolution, low frame rate video (480x270@15 fps) and the right a low resolution, high frame rate video (320x180@30 fps). Even though the amount of data applied to each pixel in both videos are about the same, the video on the left looks better when scaled to full screen.

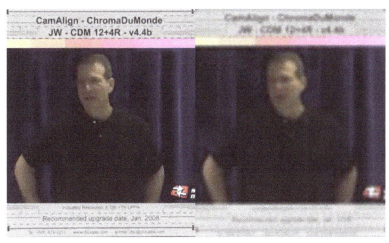

Figure 6-1. This shows the letterboxed video fit into a 1280x720 window.

For talking head videos, it's almost always better to drop the frame rate rather than resolution. Perhaps for high motion videos, the result may not be the same, so you should run your own tests.

When you need to drop the frame rate of your source videos, use the -r command. For example, here's the command I would use to convert 24 fps source to 12 fps at the lower resolution.

```
ffmpeg.exe -i TOS_1080p.mp4 -c:v libx264 -s 480x270 -r 12 TOS_480p12.mp4
```

Batch 6-1. Changing the frame rate.

Note that you can also use formulas for numbers. Here, 23.976 would become 24000/1001, 29.97 would become 30000/1001, 24 fps would be 24/1, and progressive PAL would become 25/1.

> *Note: This command adjusts the number of frames in a file that's played back at the same speed as the original. If you need to slow down or speed up a video, check bit.ly/FF_slow.*

Note: Frame rate conversions, like from 29.97fps to 24fps are beyond the scope of this book. While there are some approaches you can take with FFmpeg (see bit.ly/ff_fr1 and bit.ly/ff_fr2) most professional producers use GUI-based tools like Cinnafilm Tachyon (cinnafilm.com/tachyon/) for these operations. For a useful overview of the issues, considerations, and approaches to frame rate conversion check out Larry Jordan's article entitled, Frame Rates are Tricky Beasts (bit.ly/ff_fr3).

Other Considerations in the HLS Specification

Note that the HLS Authoring Spec referred to around Table 6-1 also contains two recommendations that I don't agree with. Specifically, the table directs producers to use same as source for the frame rate of the five highest quality variants. So, if your source is 60p, you should encode at 60p. In addition, if you're working with interlaced content, the specification says, "1.7. You SHOULD de-interlace 30i content to 60p instead of 30p."

Neither of these makes sense to me, and I detail why in a *Streaming Learning Center* article entitled, Apple Makes Sweeping Changes to HLS Encoding Recommendations (bit.ly/HLS_framerate). The CliffsNotes version is that in both cases, encoding at 60 fps can degrade the spatial quality of the video. So, I would stick to 24 fps or 30 fps.

So, we've set our basic file parameters, bitrate, resolution, and frame rate. Now, it's time to start looking at compression-related considerations, starting with keyframe and B-frame intervals.

Chapter 7: I-, B-, P-, and Reference Frames

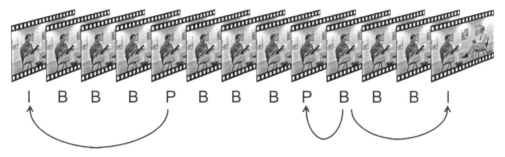

Figure 7-1. I-, B-, and P-frames illustrated.

Most codecs use different frame types during encoding. Most advanced codecs—like HEVC and H.264—use three types: I-frames (also called keyframes), B-frames, and P-frames. Figure 7-1 shows all three frame types in a group of pictures (GOP), or a sequence of frames that starts with a keyframe and includes all frames up to, but not including, the next keyframe. In this chapter, you'll learn all about these frames types, specifically:

- what I-frames are and why they're important

- I-frame configurations for single file and adaptive streaming

- what B-frames are and why they're important

- recommended B-frame configurations

- what reference frames are and why they're important

- recommended reference frame configurations

- how to configure I-frames, B-frames, and reference frames in FFmpeg.

Frame Overview

Briefly, an I-frame is entirely self-contained and is compressed solely with intra-frame encoding techniques—typically a technology like JPEG, which is used for still images on the web and in many digital cameras. P- and B-frames reference redundant information contained in other frames as much as possible. As you can see in Figure 7-1, P-frames, for predictive coded picture, can look backward for these redundancies, while B-frames, for bipredictive coded picture, can look backward and forward. This doubles the chance that the B-frame will find redundancies, making it the most efficient frame in the GOP.

Working with I-frames

How do you use these frame types to your advantage? I-frames are the largest frames, which makes them the least efficient from a compression standpoint. Basically, you only want I-frames where they enhance either quality or interactivity, or where they're mandated by your adaptive streaming segment size. This means that you deploy I-frames differently depending upon whether you're encoding for a single file or adaptive streaming.

As mentioned above, I-frames (also called keyframes) control the size of the Group of Pictures, or GOP size. So, if you have a keyframe every 60 frames in a 30 fps file, your GOP size would be 60, or two seconds for those time-based descriptions.

I-frames and Single Files

In all cases, video playback must start on an I-frame, since that's the only complete frame. When encoding a single file, as opposed to multiple files for adaptive streaming, often this file is stored on the viewer's hard drive and watched interactively, so the viewer can drag the play head slider to various points in the video. To make sure playback is responsive, when encoding a single file, you should insert an I-frame every 10 seconds, or every 300 frames in a 29.97 fps file.

How much difference in quality does I-frame interval make? Table 7-1 shows the peak signal-to-noise ratio (PSNR) ratings for 720p files, with the highest values in green and lowest in red.

	.5 Sec	1 Sec	2 Sec	3 Sec	5 Sec	10 Sec	Max Delta
Tears of Steel	38.22	39.05	39.49	39.64	39.74	39.87	4.32%
Sintel	37.09	38.06	38.57	38.75	38.97	39.08	5.37%
Big Buck Bunny	37.03	37.93	38.52	38.68	38.64	39.09	5.57%
Talking Head	43.63	44.10	44.40	44.51	44.61	44.68	2.42%
Freedom	40.33	40.67	40.88	40.96	40.99	41.03	1.72%
Haunted	41.89	42.20	42.35	42.39	42.45	42.49	1.44%
Average	**39.26**	**39.96**	**40.37**	**40.51**	**40.59**	**40.75**	**3.88%**
Screencam	35.35	38.13	37.68	38.86	40.78	41.26	16.71%
Tutorial	38.26	43.06	43.61	44.65	46.15	47.89	25.17%

Table 7-1. The impact of I-frame interval on video quality.

As the color coding shows, the longest keyframe interval produced the highest quality in all test clips, although the difference was much more significant in the Screencam and Tutorial clips than in any other. What does all this tell you? When encoding files for single file playback, use a keyframe interval of 10 or even 20 seconds, particularly for synthetic files. If you're using a keyframe interval of .5 seconds, or even 1 second, you're definitely leaving some quality on the table.

I-frames and Scene Change Detection

I-frames also improve quality when inserted at a scene change, because all subsequent P- and B-frames can reference this high-quality frame. So, when encoding for single-file delivery, you also want an I-frame at scene changes. You'll learn how to do this in FFmpeg below.

I-frames and Adaptive Streaming

The rules change completely when choosing an I-frame interval for adaptive streaming. Briefly, as you learned in Chapter 1, adaptive bitrate (ABR) technologies produce multiple streams with different quality levels to distribute to devices with varying connection speeds and CPU power. Then each stream is divided into multiple segments (also called chunks or fragments) of identical duration. During playback, the player will often choose segments from different streams to adjust to changing playback conditions, which is shown in Figure 7-2. To enable playback of each segment, you need an I-frame at the start.

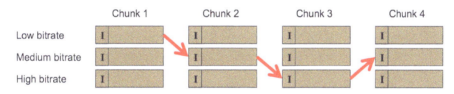

Figure 7-2. With ABR streaming, you need an I-frame at the start of each segment.

When producing ABR streams, follow these rules.

- ***The I-frame interval must divide evenly into the segment size.*** For example, in their HLS Authoring Specification, Apple mandates a segment length of six seconds, and an I-frame interval of two seconds (bit.ly/A_Devices_Spec). Though you can vary from this for DASH and other technologies, you must make sure that your I-frame interval divides evenly into your segment duration.

- ***Either disable scene change detection or force keyframes at the specified interval.*** Otherwise, you may not have an I-frame at the start of every segment.

Quick Summary: I-Frames

1. When encoding for single-file delivery, use an I-frame interval of between 10 and 20 seconds, with scene detection enabled.

2. When encoding for adaptive streaming:

- make sure the I-frame interval divides evenly into your segment size.

- disable scene change detection or force a keyframe at the required interval.

Setting I-frame Interval in FFmpeg

As you learned earlier, the I-frame controls the group of pictures (GOP) size of the video file, so an I-frame setting of 90 means a GOP size of 90. I bring this up because the easiest way to remember FFmpeg's I-frame switch is to think G for GOP size. When setting the I-frame interval in FFmpeg, you have the three controls in Batch 7-1 to consider.

```
ffmpeg -i TOS_1080p.mov -c:v libx264 -g 48 -keyint_min 48 -sc_threshold 0 TOS_G.mp4
```

Batch 7-1. Keyframes at regular intervals and not at scene changes.

Here's a description of what the switches do.

`-g 48` sets the maximum keyframe interval at 48, or every two seconds for a 24 fps file. If not set, FFmpeg will insert an I-frame every 250 frames. For single files, you can use an interval of five to 10 seconds, or just don't set this option. For files produced for ABR streaming, use an interval that divides evenly into your segment size, usually either two or three seconds.

`-keyint_min 48` is the minimum distance between I-frames. If not set, FFmpeg will use a minimum interval of 25. When encoding a single file, you can simply not set this option. When producing for adaptive streaming, you should set the minimum to the same value as the GOP size to ensure I-frames at the specified interval.

`-sc_threshold 0` sets the threshold for scene detection. If not set, the threshold is 40. This is fine for single files, but not adaptive. When producing for adaptive streaming, you should disable scene change detection with the setting of 0 as shown in the string.

These commands deliver the keyframe interval shown in Figure 7-3.

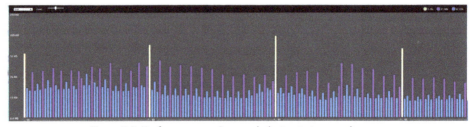

Figure 7-3. Keyframes every 2 seconds, but not at scene changes.

Tip: *The tool in Figure 7-3 is Telestream Switch. It's a great tool for identifying frame tapes (I and B) in H.264 and HEVC files, but you'll need to buy Switch Pro ($295) to get this view (primary.telestream.net/switch/.*

Inserting I-frames at Specified Intervals and Scene Changes

To insert I-frames at the specified interval *and* at scene changes, use the switches shown in Batch 7-2.

```
ffmpeg -i TOS_1080p.mov -c:v libx264 -force_key_frames expr:gte(t,n_forced*2)
-keyint_min 25 -sc_threshold 40  TOS_scene.mp4
```

Batch 7-2. Keyframes every two seconds and at scene changes.

Here's a description of what the switches do.

> `-force_key_frames expr:gte(t,n_forced*2) -keyint_min 25`
> `-sc_threshold 40`

`-force_key_frames expr:gte(t,n_forced*2)` forces the I-frame at the keyframe interval of two seconds. Substitute the desired I-frame interval in seconds for the 2 in the string.

`-keyint_min 25` sets the minimum I-frame interval at 25, which is the default. If the default value is acceptable, don't include this switch.

`-sc_threshold 40` sets the threshold for scene detection at 40, which is the default. Again, if the default value is acceptable, don't include this switch.

This produces the keyframe cadence shown in Figure 7-4. With a keyframe interval of 2 or 3, my tests reveal that keyframes at scene changes add very little extra quality. For example, the file shown in Figure 7-3 had a PSNR value of 41.22207 dB, while the file in Figure 7-4 had a PSNR value of 41.25565 dB, about 0.08 percent higher. So, I tend to keep it simple and not use keyframes at scene changes, but it's your choice.

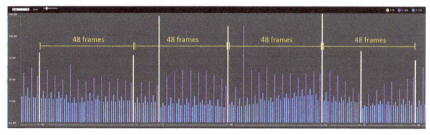

Figure 7-4. Keyframes at regular intervals, and at scene changes.

Tip: *There's a lengthy discussion of the best option for achieving consistent keyframe intervals for Dynamic Adaptive Streaming over HTTP (DASH) encoding at bit.ly/Iframes_DASH. The recommendations shown in Table 7-8 follow the recommendation there, but it's not the only way to skin this particular cat.*

Working with B-frames

Now let's turn our attention to B-frames. As we discussed in Chapter 3, the B-frame interval is controlled by the encoding preset. If you don't specify a preset, FFmpeg will use the Medium preset, which means a B-frame interval of eight. This means that FFmpeg will attempt to insert up to eight B-frames between P-frames or I-frames and P-frames.

I say up to because by default, FFmpeg uses adaptive B-frame placement, and only inserts B-frames when it improves quality. This means that even if you specify a B-frame interval of 3, the number of B-frames between P-frames and I-frames will vary, sometimes three, sometimes less than three, but never more than three.

What B-frame interval delivers the highest quality? As you can see in Table 7-2, an interval of three delivers the highest overall quality, though the difference is exceptionally slight between an interval of three and all other settings. So, if you want to simply go with the default value, you're not going to experience a significant loss.

	0B	1B	2B	3B	4B	5B	10B	15B	Max Delta
Tears of Steel	38.95	39.41	39.56	39.65	39.62	39.61	39.60	39.63	1.75%
Sintel	38.34	38.71	38.74	38.76	38.76	38.75	38.75	38.75	1.07%
Big Buck Bunny	39.96	40.34	40.41	40.40	40.38	40.41	40.40	40.39	1.13%
Talking Head	44.21	44.44	44.50	44.52	44.51	44.51	44.50	44.50	0.68%
Freedom	40.76	40.93	40.93	40.96	40.93	40.93	40.91	40.91	0.49%
Haunted	42.19	42.33	42.39	42.41	42.36	42.38	42.36	42.36	0.50%
Average	40.74	41.03	41.09	41.11	41.09	41.10	41.09	41.09	0.94%

Table 7-2. The impact of B-frame configurations on quality as measured by PSNR.

Quick Summary: B-Frames

1. Always use B-frames when available.

2. Since the actual number of B-frames doesn't matter significantly, you typically can use the number specified by the x264 preset used on the command line argument.

Inserting B-frames in FFmpeg

When you set B-frames with FFmpeg, you should set these two values.

`-bf 3` sets the desired value for B-frames, in this case, the recommended interval of 3. Note that if you don't manually insert a B-frame interval, but do insert a preset (like Very Slow), the preset controls the B-frame interval. If you don't insert an interval or choose a preset, I believe FFmpeg uses 16 but I was unable to confirm this.

`-b_strategy 2` enables adaptive B-frame placement, and there are three option. 0

is very fast, but not recommended. 1 is the faster and default mode, and 2 is the slower mode that I recommend. You can read about why at bit.ly/OptimizeBs. To make a long story short, choosing 2 instead of 1 roughly doubled the number of B-frames inserted into the file. If you don't choose a setting, but choose a preset, the setting in the preset controls this setting. If you don't choose a preset or a specific setting, FFmpeg uses a value of 1.

Figure 7-5 shows how much difference the second switch makes in terms of B-frames actually inserted into the file. On top, with a setting of 1, only 9 percent of the frames are B-frames. With a setting of 2, 58 percent of the frames are B-frames.

Figure 7-5. A setting of 2 inserts many more B-frames into the file but doesn't significantly improve quality.

Interestingly, despite the second file having more than six times the number of B-frames, the quality improvement was only .01% higher.

Reference Frames

Briefly, a reference frame is a frame that the frame being encoded can use for redundant information. As with B-frames, if you don't explicitly set a reference frame value, FFmpeg will use the value set by the default preset. If you don't specify a preset, FFmpeg uses the Medium preset, which means a reference frame value of 3 frames.

What's the best value? Intuitively, increasing the number of reference frames will increase encoding time, because the encoder must search more frames for redundancies. For example, set reference frames at 1, and once the encoder searches a single frame with redundancies, it's done. Set it at 16, and the encoder must search through 16 frames. The hoped-for benefit is an increase in quality, as more redundancies should translate to higher quality.

Let's look at the quality side first. Table 7-3 explores this question, with 720p files encoded using a B-frame interval of 3. As always, the red columns mean the lowest value, the green value the highest. As you can see, with all videos except Tears of Steel, 1 was always the lowest score, while 16 averaged the highest score, though the Max Delta quality differences were very minor.

Note that the difference between 10 and 16 reference frames in the top six videos averaged a minuscule 0.03 percentage points.

Average Quality	1 Ref	5 Ref	10 Ref	16 Ref	Max Delta	10 - 16 Delta	16 - 5 Delta
Tears of Steel	39.34	38.99	39.47	39.49	1.28%	-0.04%	-1.26%
Sintel	38.45	38.54	38.58	38.59	0.35%	-0.02%	-0.12%
Big Buck Bunny	39.99	40.09	40.11	40.11	0.31%	0.00%	-0.05%
Talking Head	44.27	44.36	44.39	44.40	0.29%	-0.03%	-0.10%
Freedom	40.68	40.80	40.85	40.87	0.47%	-0.06%	-0.19%
Haunted	42.24	42.32	42.35	42.36	0.26%	-0.02%	-0.08%
Average - 720p	40.83	40.85	40.96	40.97	0.34%	-0.03%	-0.29%

Table 7-3. The impact of the number of reference frames on PSNR quality.

Reference Frames and Encoding Time

Regarding encoding time, Table 7-4 tells the tale. As you can see, the average encoding time delta between 1 or 16 reference frames was 136 percent. Dropping down to 10 reference frames shaved 21 percent off encoding time, while dropping to 5 reference frames reduced encoding time by 43 percent.

Encoding Time	1 Ref	5 Ref	10 Ref	16 Ref	Max Delta	10 - 16 Delta	16 - 5 Delta
Tears of Steel	39	49	72	91	133%	-21%	-46%
Sintel	40	53	71	76	90%	-7%	-30%
Big Buck Bunny	41	53	68	85	107%	-20%	-38%
Talking Head	37	47	61	77	108%	-21%	-39%
Freedom	99	142	200	263	166%	-24%	-46%
Haunted	47	65	93	123	162%	-24%	-47%
Average - 720p	51	68	94	119	136%	-21%	-43%

Table 7-4. Reference frames and encoding times.

To put this in perspective, if you're currently encoding using 16 reference frames, you can cut your encoding time roughly in half by switching to 5 reference frames (from 119 seconds to 68). The toll on quality? A drop from 40.70 to 40.58 dB, or about 0.3 percent. Or, you can drop your reference frames to 1, save a bunch more time, and lose only about 0.41 percent in PSNR quality.

Quick Summary: Reference Frames

1. The qualitative difference relating to reference frames is insignificant. For most users, this makes the reference frame setting an encoding time issue rather than a quality issue. For the record, unless encoding time is critical, I typically recommend a setting of 10.

2. If you're operating at capacity, and are about to buy new gear, you can cut encoding time significantly by dropping the number of reference frames down to 1. The total quality delta will be less than 0.5 percent.

Reference Frames in FFmpeg

To set reference frames in FFmpeg, use the `-refs` switch.

`-refs 10` inserts that input value, in this case, 10, for reference frames.

Pulling this all together, a simple one-pass encode might look like this.

```
ffmpeg -i TOS_1080p.mov -c:v libx264 -g 48 -keyint_min 48 -sc_threshold 0
-bf 3 -b_strategy 2 -refs 10 TOS_frames.mp4
```

Batch 7-3. Controlling I-, B-, P- and reference frames.

This produces an MP4 file with a keyframe interval of two seconds, with no keyframes at scene changes. FFmpeg will insert 3 B-frames using the second strategy, and will search ten reference frames for each encoded frame. Audio would pass through unchanged from the source. Note in Figure 7-6 that the actual B-frame interval varies from 1-3, as discussed previously.

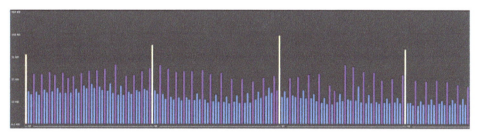

Figure 7-6. Regular I-frames, but B-frame interval varies.

That's it for frame types. In the next chapter, you'll learn all about the H.264 standard and codec, the closest we have to a one-size-fits-all codec.

Chapter 8: Encoding H.264

H.264 is as close to a universal codec as there is today, playing on all computers, mobile devices, Smart TVs, set-top boxes, OTT devices, and pretty much anything else that can play a video file. In previous chapters, we performed all tests using the H.264 codec. Most of those lessons were generic and apply equally to other codecs. In this chapter, we'll discuss H.264-specific configuration options. Specifically, you will learn:

- what profiles and levels are and how they affect quality and playback compatibility

- what context-adaptive binary arithmetic coding (CABAC) and context-adaptive variable-length coding (CAVLC) are and when to choose one over the other

- how x264 presets and tuning mechanisms operate and how to use them.

What Is H.264?

H.264 is a video compression technology, or codec, that was jointly developed by the International Telecommunication Union (as H.264) and International Organization for Standardization/International Electrotechnical Commission Moving Picture Experts Group (as MPEG-4 Part 10, Advanced Video Coding, or AVC). Thus, the terms "H.264" and "AVC" mean the same thing and are interchangeable.

Note that there are many H.264 codecs, all with different strengths and weaknesses. All should create streams compatible with all H.264 players. The H.264 codec in FFmpeg is x264, which is widely considered to be the best available H.264 codec, which I've found to be true in my tests. But, you should be aware that it's not the only H.264 codec out there.

For the record, all H.264 codecs can use different profiles and levels, and with CABAC or CAVLC enabled. Only x264 uses the presets and tuning mechanisms covered in this chapter, though most other H.264 codecs have similar presets that trade off quality for encoding speed.

MP4 Container Format

As defined back in Chapter 1, a container format is "a meta-file format whose specification describes how data and metadata are stored" (bit.ly/containerformat). When encoding with the H.264 codec, you must choose the correct container format for your target player; otherwise, the file won't play. For example, even though iPhones can play .ts files containing video encoded with the H.264 codec, they can't play .ismv files encoded with H.264 but packaged into the Smooth Streaming container format. I'll cover the container formats you should use for each ABR format in later chapters.

If you had to choose one container format to produce a single file that plays almost everywhere, use the MPEG-4 container format and produce an .mp4 file. That's because all players—whether Flash, Silverlight, iPad, iPod, Android, or Windows Phone—play .mp4 files. Most adaptive streaming formats use different containers, which you'll have to customize for each target.

Note that the H.264-specific encoding options you'll learn in this chapter apply to all H.264-encoded files, irrespective of the intended container. So, while you'll have to change your container for each target, the H.264-specific encoding parameters themselves remain the same.

Basic H.264 Encoding Parameters

Let's start with the basics—profiles and levels—which are the most fundamental configuration options you can access from almost all encoding programs, and of course, FFmpeg.

Profiles and Levels

Profiles and levels are the most basic H.264 encoding parameters and are available in most H.264 encoding tools. According to the now changed (but less descriptive) Wikipedia definition, (en.wikipedia.org/wiki/H264), a profile "defines a set of coding tools or algorithms that can be used in generating a conforming bitstream," whereas a level "places constraints on certain key parameters of the bitstream." In other words, a profile defines specific encoding techniques that you can or can't use when encoding a file (such as B-frames), while the level defines details such as the maximum resolutions and data rates within each profile.

Why do profiles exist? To define different grades of H.264 that can be used by different devices depending upon the power of their CPU. For example, the original iPods and iPhones could only play the Baseline profile, so video encoded using the Main or High profiles won't play on these devices. In contrast, most computers and all over-the-top (OTT) devices like Apple TV and Roku boxes can play video encoded using the High profile, as well as video encoded using the Baseline or Main profile.

When choosing a profile, compatibility is the key issue, and it's really only an issue for older mobile devices. Through 2016, Apple recommended that lower rungs on the encoding ladder use the Baseline profile to maintain compatibility with older iDevices. Then, in the HLS Authoring Specification for Apple Devices (bit.ly/A_Devices_Spec), Apple states, "1.2. Profile and Level MUST be less than or equal to High Profile, Level 4.2," and "1.3. You SHOULD use High Profile in preference to Main or Baseline Profile." So, you're not required to use the High profile, but it's recommended.

Unfortunately, the Android scenario isn't quite so clear, because Google doesn't know which profile the multitude of Android devices support in hardware. The Android operating system itself supports only the Baseline profile (bit.ly/androidvideospecs) and that's what Google recommends for the broadest possible compatibility.

Life would be easiest for all streaming professionals if they could produce one set of files that played everywhere, but that would have to use the Baseline profile. How much quality would you leave on the table? Well, let's have a look.

Comparative Quality—Baseline, Main, and High Profiles

Table 8-1 shows the comparative quality of 720p videos encoded using the Baseline, Main, and High profiles. As you can see, the Baseline profile delivers the lowest quality in all cases, and the High profile the best. What might be surprising is how little difference there is; only an average of 2.84 percent for all clips, with about 80 percent of that between the Baseline and Main profiles. The difference between Main and High is only half a percent.

Average Quality	Baseline	Main	High	Delta - Baseline/Main	Delta - Main/High	Total Delta
Tears of Steel	37.52	39.11	39.46	4.26%	0.88%	5.19%
Sintel	37.13	38.27	38.58	3.08%	0.78%	3.90%
Big Buck Bunny	38.45	39.82	40.11	3.56%	0.72%	4.31%
Talking Head	43.69	44.34	44.39	1.48%	0.12%	1.60%
Freedom	39.60	40.62	40.85	2.57%	0.58%	3.17%
Haunted	41.55	42.22	42.35	1.60%	0.31%	1.91%
Screencam	46.43	46.74	46.87	0.67%	0.28%	0.95%
Tutorial	43.44	44.04	44.18	1.38%	0.32%	1.71%
Average	40.98	41.89	42.10	2.32%	0.50%	2.84%

Table 8-1. Comparative quality of the different profiles at 720p.

So, in many instances—particularly at higher quality levels, like our 720p files—the difference in overall quality between the Baseline, Main, and High profiles likely wouldn't be distinguishable by the average viewer. Here's a quick summary on profiles.

Quick Summary: Profiles

1. In general, use the highest-quality profile your target device will support.

2. That said, the quality difference between the profiles isn't as much as you might think. When encoding for multiple platforms, it may make sense to encode with the Baseline profile to create one file (or one set of files) that you can deliver to all targets.

Choosing Profiles in FFmpeg

The default profile used by FFmpeg depends upon the compilation; in the tests shown in Chapter 1, FFmpeg defaulted to the High profile. To change that, insert the following string:

```
-profile:v baseline or main
```

H.264 Levels

What about H.264 levels? As mentioned earlier, levels provide bitrate, frame rate and resolution constraints within the different profiles. Whenever you encode a file, the encoder inserts a level into the file metadata. Before attempting to play the file, devices will check the metadata to make sure the level doesn't exceed its capabilities.

In this role, levels enable primarily device vendors to further specify the types of streams that will play on their devices. For example, the iPhone SE Apple released in the summer of 2016 will play, "H.264 video up to 4K, 30 fps, High Profile level 4.2 with AAC-LC audio up to 160 kbps, 48 kHz, stereo audio in .m4v, .mp4, and .mov file formats."

Accordingly, when you're producing for devices, you need to ensure that your encoding parameters don't exceed the specified level, which again should be designated by the device manufacturer. Otherwise, the file may not play smoothly or even load.

Levels and Computers/OTT

Where levels are critical for mobile playback, they are irrelevant when encoding for computer playback because software-based streaming players, like Adobe Flash and the key browsers for HTML5 playback, can play H.264 video encoded using any profile or any level. Ditto for smart TVs and OTT boxes like Roku or Apple TV.

Quick Summary: Levels

1. It's critical to ensure that streams bound for older mobile devices don't exceed the specified levels.

2. Levels are typically not relevant when encoding for computers, smart TVs, or OTT devices, where the resolution and data rate of your video are the most important considerations.

Setting Levels in FFmpeg

I'm not sure how FFmpeg chooses the default level for its encodes. In every case I checked, the level inserted by FFmpeg exceeded the level that seemed appropriate for the actual file parameters. For example, I encoded a file at 640x360@600 kbps using the High profile, which according to my reading of Wikipedia, was within the specs for Level 3. FFmpeg injected Level 3.1 into the file metadata. I next encoded a file at 1280x720@6000 kbps using the High profile, which fits under the constraints of Level 3.1, and FFmpeg injected Level 5 into the file metadata. It's possible that there were other constraints coming into play that forced FFmpeg into the higher level, like macroblocks or luma samples, but I couldn't tell this from the files.

Of course, if FFmpeg is inserting too high a level, that may cause those files to be rejected by a target device that could play the file if the level designation was correct. So, if you're using FFmpeg to create files for mobile playback, you better sort this out by manually inserting the correct level and testing playback on various target devices.

You set levels in FFmpeg using the following string.

```
-level:v <integer>
```

So, if you wanted to encode to Level 2.2, you would insert the string `-level:v 2.2` anywhere in the command line.

Note that inserting the level into the command string does not force FFmpeg to constrain the encoding parameters to those specified by the level specifications. For example, I encoded a file to 1280x720@6000 kbps using the High profile, and inserted Level 2.2 into the command string as shown above. According to Wikipedia, Level 2.2 tops out at 720x276@12.5 fps. As shown in Figure 8-1, FFmpeg produced the file at the original target parameters, which far exceed those set in the levels.

Figure 8-1. FFmpeg does not constrain file encoding to the designated level.

The bottom line is that if you're using FFmpeg to produce for mobile devices, you need to make sure that your file parameters don't exceed the level designated for each device and that you properly insert the level into the command string.

Entropy Coding

When you select the Main or High profiles, you'll have two options for entropy coding, which controls how the frame-related compressed data is packed before storage. The options are CAVLC, for context-based adaptive variable length coding, and CABAC, for context-based adaptive binary arithmetic coding. Of the two, CAVLC is the lower-quality, easier-to-decode option, while CABAC is the higher-quality, harder-to-decode option.

The essential point is that CABAC delivers a small quality improvement with very little increased CPU requirements upon playback. So, when encoding with the Main or High profiles, you should always use CABAC.

Quick Summary: Entropy Coding

> ***1.*** CABAC delivers a small quality improvement with very little increased CPU requirements upon playback
>
> ***2.*** CABAC should be enabled whenever encoding with the Main and High profiles (it's not available with the Baseline profile).

Setting Entropy Encoding in FFmpeg

The default entropy encoding setting for FFmpeg is CABAC when it's available, which is the Main and High profiles. FFmpeg uses CAVLC for the Baseline profile, which is the only option. If, for some reason, you wish to use CAVLC when encoding in the Main or High profile, use the following string:

```
-coder 0
```

Substituting 1 for 0 forces CABAC (`-coder 1`), but you can achieve the same result by omitting the -coder command string entirely and going with the default.

x264 Presets and Tuning

So far, in previous chapters and this one, we've learned how to set most of the basic options available in most x264-based encoding tools. However, we're just touching the surface of the full range of options in the codec. For example, there are configuration options that control the range of the search that the encoder performs when looking for redundancies, and the precision of that search. There are literally dozens of other controls and attempting to identify their best settings for the different types of videos covered in this book would be daunting.

Fortunately, with the x264 codec, you don't have to. Rather, the developers of the codec have created presets that adjust many of the x264 configuration options to trade off encoding time with quality. These are shown on the left in Figure 8-2 (from Ultra Fast to Placebo). On the right are tuning mechanisms that enable producers to customize their encodes for certain types of videos (film, animation, still image), for low latency or fast decode, or to perform well on different quality benchmarks (SSIM, PSNR). Rather than study the individual configuration options available in x264, we'll focus on these presets and tuning mechanisms.

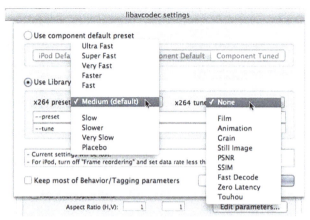

Figure 8-2. x264's presets and tuning mechanisms.

x264 Presets

There are 10 x264 presets, ranging from the low-quality/high-speed Ultrafast, to the optimum-quality/slowest-speed Placebo. These presets are available only for the x264 codec, so they won't be available if you're working with MainConcept or a different H.264 codec.

Most encoders let you choose these presets via controls like those shown in Figure 8-2, and you'll learn how to choose the presets in FFmpeg in a moment. Looking ahead, as you'll see in Chapter 12, the x265 codec also uses presets with the same names as those used in x264.

I'm not going to detail the options in each preset since they're very well defined on various websites (like bit.ly/x265_preset_details). Rather, I'm going to focus on quality and encoding speed, starting with the quality side as shown in Table 8-2. To produce the table, I encoded all videos at the 720p configurations using the High profile.

Average Quality	Ultrafast	Superfast	Veryfast	Faster	Fast	Medium	Slow	Slower	Veryslow	Placebo	Total Delta
Tears of Steel	36.07	37.82	38.51	39.23	39.26	39.33	39.27	39.41	39.47	39.40	9.43%
Sintel	35.14	36.71	37.42	38.40	38.43	38.46	38.40	38.55	38.57	38.47	9.75%
Big Buck Bunny	36.23	38.01	38.92	39.97	40.02	40.03	40.01	40.12	40.12	40.06	10.74%
Talking Head	43.38	43.38	44.06	44.39	44.28	44.28	44.21	44.34	44.39	44.29	2.34%
Freedom	38.46	39.26	40.01	40.41	40.32	40.58	40.55	40.69	40.85	40.77	6.22%
Haunted	41.13	41.30	41.89	42.20	42.07	42.27	42.25	42.27	42.35	42.31	2.98%
Screencam	44.46	45.67	46.68	47.12	46.82	46.96	46.95	47.06	46.88	46.76	5.99%
Tutorial	38.47	41.83	43.62	44.50	44.37	44.30	43.99	44.14	44.07	43.91	15.68%
Average	38.40	39.41	40.13	40.77	40.73	40.83	40.78	40.90	40.96	40.88	7.89%

Table 8-2. Output quality in PSNR value by x264 preset.

As you can see, the Ultrafast preset consistently produced the lowest-quality output, with the best quality strangely split between Faster and Veryslow. The difference between the best and worst values averaged only 7.89 percent, although it did range as high as 15.68 percent for PowerPoint-based tutorial footage.

What about encoding time? This is shown in Figure 8-3, where you see that encoding time stays fairly low through the Veryfast preset, then starts to extend, but doesn't really blow up until the Placebo preset.

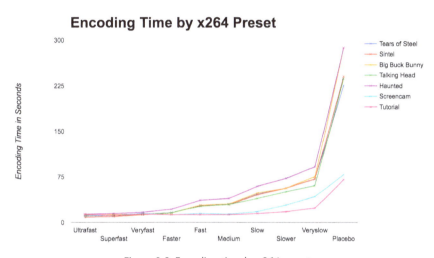

Figure 8-3. Encoding time by x264 preset.

How to synthesize this data? Well, I help you do just that in Figure 8-4. To explain, the figure:

- doesn't include the Tutorial and Screencam videos, because they perform so much differently than the others

- averages the quality of all videos and presents quality as a percentage of the maximum. So, the lowest score (Ultrafast) is 93.8% of total quality, while the highest score (Veryslow) is 100 percent

- averages the encoding time of all presets and presents the average as the percentage of the difference between the fastest and slowest encoding time. So, the fastest time is 0 percent, while the slowest time is 100 percent.

Essentially, the chart illustrates the encoding time/quality trade-off. As an example, the medium preset took 8.6 percent of the encoding time of the placebo preset, but delivered 99.7 percent of the quality. What information can we derive from this chart?

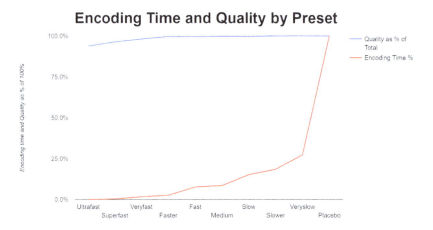

Encoding Time and Quality by Preset

Figure 8-4. Plotting quality against encoding time with real-world videos and animations.

• First, that the Faster preset is the minimum preset that you should run for production videos, capturing 99.5 percent of the available quality in 2.6 percent of the encoding time.

• At the other end of the spectrum, if encoding time isn't an issue, you should use the Veryslow preset, which delivers 100 percent of the available quality in 27.3 percent of the time of the Placebo preset.

• Finally, you should never use the Placebo preset, since it delivers less than 100 percent of the available quality while taking three times longer than the Veryslow preset.

Tip: *One of the more impressive analyses of the x264 presets is available at bit.ly/xpreset. One of the observations is that the choice of preset makes a more significant difference at lower data rates. Overall, the author concludes, "if we have more time for the compression and if the sharp and detailed image are important for us, slow and slower presets are, in my opinion, the best ones (with an emphasis on the latter). … If the speed of compression (not the image quality) is a priority, I propose to choose faster preset (surprisingly good results for SSIM and PSNR indexes) or fast."*

Quick Summary: x264 Presets

1. x264 Presets are a convenient way to choose a range of configuration options that trade off video quality against encoding time.

2. For real-world videos, the Faster preset is probably the fastest option that you should select, since it captures 99.5 percent of the available quality in 2.6 percent of the encoding time. If encoding time isn't an issue, use the Veryslow preset, which delivers 100 percent of the available quality in 27.3 percent of the time of the Placebo preset.

3. You should never use the Placebo preset.

Choosing an x264 Preset

If you don't choose a preset, FFmpeg essentially applies the parameters of the medium preset, which is the default in most programs that use x264. To choose a different preset, insert the following string into the command line:

```
-preset <preset name>
```

So, you would choose the Fast preset with the string `-preset fast`.

Tuning Mechanisms

The last subject we'll cover in this chapter is tuning mechanisms, again designed for certain types of videos or to produce certain types of performance. We'll look at three tuning mechanisms: animation for animated videos, and film and grain for film-based videos (or those supposed to look like them).

Like presets, tuning options are available only with the x264 codec, not other H.264 codecs, and they aren't presented in all encoding programs that support x264. Most desktop programs that enable x264 tuning do so as shown in Figure 8-2, where you simply select the tuning mechanism in the program's user interface. You'll learn how to select tuning options in FFmpeg at the end of this chapter.

Again, I'm not going to delve into the details of each tuning mechanism as there are multiple sources that do so (see bit.ly/x264_tune for one). Rather, I encoded with and without the tuning mechanism to learn whether it increased quality and if so, by how much.

Animation Tuning

If you produce a lot of animated videos with the x264 codec, the animation tuning mechanism looks like a winner. You can see this in Table 8-3, where I supplemented the Big Buck Bunny and Sintel animations with two *SpongeBob SquarePants* movie trailers, and a short animated clip from a cartoon called El Ultimo. Specifically, the first three columns show the clips encoded with the High profile, then the High profile with animation tuning. As you can see, the clips encoded with animation tuning averaged 1.66 percent higher.

Presets - Animation	PSNR			Low Frame		
	High	Anim	Delta	High	Anim	Delta
Big Buck Bunny	39.64	40.08	1.11%	24.75	25.11	1.44%
El Ultimo	45.85	46.58	1.59%	41.64	41.88	0.56%
Sintel	38.57	39.11	1.39%	28.85	29.53	2.35%
SpongeBob 1	37.46	38.21	1.99%	26.10	26.21	0.41%
SpongeBob 2	39.29	40.17	2.24%	33.41	33.53	0.34%
Averages	40.16	40.83	1.66%	30.95	31.25	1.02%

Table 8-3. The positive benefits of tuning for animation.

I've recommended the animation tuning mechanism to multiple clients who stream animated content, and you should try it as well. Interestingly, I ran the same tests with the other clips we've been working with, and animation tuning improved their quality as well, boosting the quality of the Freedom clip by 2.39 percent. I'm hesitant to recommend animation tuning for all clips because the configuration options are specifically designed for animated content. But it's worth a try if you have the time to subjectively confirm the PSNR results.

Film and Grain Tuning

The results for film and grain tuning were not as positive. To explain, the first set of columns in Table 8-4 show three movie clips encoded using the High profile, and then the High profile with film tuning. Here, film tuning dropped quality by 0.15 percent. The second set of three columns explores the grain tuning mechanism on the same three clips. Here, the benefit was minuscule at 0.12 percent.

	High	Film	Delta	High	Grain	Delta
Elektra	42.35	42.54	0.46%	42.35	42.71	0.87%
Tears of Steel	41.47	41.32	-0.37%	41.47	41.28	-0.47%
Zoolander	40.31	40.10	-0.53%	40.31	40.29	-0.05%
Average	41.38	41.32	-0.15%	41.38	41.43	0.12%

Table 8-4. Film and grain tuning produced meh benefits.

Overall, I've never been a fan of film and grain tuning options and nothing I see here reverses that trend.

Quick Summary: x264 Tuning

1. x264 tuning mechanism are designed to improve quality or performance for specific types of videos or quality metrics.

2. Animation tuning seems to improve the quality of all animated videos as measured by PSNR. If you distribute a lot of animated content, you should experiment with this tuning mechanism to see if it improves the quality of your animations.

3. The film and grain tunes have not worked well for me in this and other tests.

Choosing an x264 Tuning Mechanism

By default, x264 doesn't apply a tuning mechanism. To deploy a tuning mechanism, use the following string:

```
-tune <tune name>
```

So, you would choose the Animation preset with the string `-tune animation`.

> ***Tip:*** *Note that if you're comparing x264 with other codecs using PSNR or SSIM computations, you should enable the PSNR and SSIM tuning algorithms. According to the codec's developers, these switches disable optimizations that improve subjective quality but can degrade PSNR and SSIM scores. You can read a post on this at bit.ly/x264_psnrtune.*

OK, now you know all you need to know to go forth and conquer when encoding with x.264 in FFmpeg. May the force be with you. We take a shorter look at audio in the next chapter.

Chapter 9: Working with Audio

Unless you're into silent movies, audio will be a major component of any production. In this chapter, you will learn how to encode the audio that accompanies your video. Specifically, you will learn:

- which audio codecs to use for different video codecs
- FFmpeg names for these audio codecs
- How to configure audio sample rate, channels, and bitrate.

Which Audio Codec?

When producing MP4 files for single file delivery in an MP4 container, use the acc codec via the `-c:a` or `-acodec` string, so designating aac would mean the following.

```
-c:a aac
```

If you don't designate aac, and encode an MP4 file, FFmpeg will default to that codec. As you can see from Table 3-1, FFmpeg will also reduce the data rate to a more reasonable level, in the case of our test files, reducing the data rate from 317 kbps to 132 kbps.

Note that there are two versions of AAC designed for low bitrate delivery, HE-AAC v1, and HE-AAC v2. I've never been a fan of these formats, and was unable to produce them with FFmpeg despite several attempts. It's almost certainly user error, but given that I don't recommend using them, I didn't want to invest additional time trying to produce them. In the HLS Authoring Specifications, Apple identifies all three AAC-based formats as compatible, but states, "You SHOULD NOT use HE-AAC if your audio bit rate is above 64 kb/s" (bit.ly/A_Devices_Spec).

Dolby Digital

When encoding for HLS delivery to Apple devices, you can also encode using one or two Dolby Digital codecs, which play natively on the latest iOS devices, as well as Windows 10. Dolby Digital (ac3), was the original Dolby Digital format that can deliver up to six channels of audio at data rates between 384 and 640 kbps. Dolby Digital Plus (eac3), evolved from Dolby Digital, and is a more capable audio format that brings higher audio quality, lower data rates, more powerful metadata-driven features and support for 7.1 audio channels.

Here are strings to choose the Dolby codecs.

`-c:a ac3` to choose Dolby Digital

`-c:a eac3` to choose Dolby Digital Plus

Note that beyond support for surround channels, Dolby Support also adds higher quality compression than AAC, plus loudness control. If you're producing premium content for HLS delivery, you should definitely consider these Dolby formats. Note that I produced a three-part video tutorial in conjunction with Dolby entitled, Best Practices for Encoding and Delivering Dolby Audio to Apple Devices (bit.ly/HLS_Dolby). If you're interested in learning more about how to add Dolby to HLS, you should definitely check it out.

Opus Audio for VPX

If you're encoding audio for VP8 or VP9 video files, you'll need to convert the audio to the Opus codec, using this designation.

`-c:a libopus`

Controlling Audio Parameters

Typically, there are three audio parameters that you need to control; bitrate, sampling rate, and channels. You control them with these switches.

Bitrate

Bitrate is the bitrate of the audio file, which you control with the following switch.

`-b:a 128k` bitrate audio, followed by the bitrate (128 kbps here).

If the bitrate of your input is 128 kbps (or so), and you're encoding with AAC, FFmpeg will pass that value through. If it's much higher, as it was back in Chapter 3, FFmpeg will probably reduce it to around 128 kbps for AAC audio, and 196 kbps for both Dolby audio formats. To make sure you get the bitrate that you want, you should specify bitrate in all command strings.

In terms of recommended bitrate, Table 9-1 shows Apple's data rate recommendations for the various codecs and configurations. Note that Apple recommends a maximum data rate for AAC of 160 kbps, which seems to indicate that values beyond this add no perceivable quality. That's my experience, though I have seen values in encoded files of 256 kbps and higher.

What about Opus? I don't have a lot of production experience with Opus, but I found several comparisons that rated Opus higher than AAC at comparable bitrates (bit.ly/opus_aac). So, you probably can feel comfortable using the same bitrate for Opus as with AAC.

Audio channels	Format	Total (kb/s)
2.0 (stereo)	AAC	32 to 160
2.0 (stereo)	Dolby Digital Plus	96 to 160
5.1 (surround)	Dolby Digital	384
5.1 (surround)	Dolby Digital Plus	192
7.1 (surround)	Dolby Digital Plus	384

Table 9-1. Recommendations from Apple's HLS Authoring Specifications.

Sample Rate

The sample rate is the number of times per second the audio is sampled, measured in hertz (Hz) or kilohertz (kHz), with higher values delivering higher fidelity. This is easiest to understand within the concept of converting analog audio to digital. That is, if you sampled a recording from the London Symphony Orchestra once per second, you'd have a string of one-second beeps and blurts that sound nothing like the original. If you sampled at 48000 Hz, you'd have 48,000 discrete samples per second, which should deliver a very accurate reproduction of the actual sound.

By default, FFmpeg will pass through the sampling rate of the input file. So long as your input rate is 48000 Hz or lower, you probably don't want to change it, so you don't need to include this control. If the sampling rate is 96000 Hz or higher, you probably do want to change it, using the following switch.

`-ar 48000` sampling rate, followed by the bitrate (48000 Hz here).

Channels

Channels are the number of discrete streams in the recording, with mono being a single stream containing all input, stereo two streams containing the left and right tracks, and 5.1 and 7.1 separate streams of surround input.

By default, FFmpeg will pass through the number of channels in your input file, so if you have stereo input and want stereo output, you don't need to include this switch. If you're working with surround input, or want to change stereo to mono, you do need to include this switch.

`-ac 2` audio channels, followed by the number (here 2, or stereo).

There's a misconception that if you produce a mono stream, the listener will only hear audio from one speaker or one side of the headphones. This isn't the case, as the player simply sends the mono stream out both audio channels.

In many instances, like talking head videos, mono is just as good as stereo, since it's recorded by either a single mic or by two mics in a camcorder that are so close together that the left and right channels are nearly identical. If you encode that audio in stereo, you have to allocate twice the bitrate to achieve the same quality as you would for a mono encode. So your stereo stream might take 128 kbps while delivering almost the exact same experience from a mono stream encoded at 64 kbps. Accordingly, for talking head audio, you should strongly consider mono.

> **Tip:** *In chapter 10, you'll learn how to produce audio-only and video-only files.*

Putting it All Together

Here's a command string that pulls together the audio-related concepts.

```
ffmpeg.exe -i TOS_1080p.mov -c:v libx264 -c:a aac -b:a 128k -ac 2 -ar 48000
TOS_1080p.mp4
```

Batch 9-1. Setting audio parameters in FFmpeg.

Here's an explanation for the commands used in this string.

`ffmpeg.exe` calls the program.

`-i TOS_1080p.mov` names the input file.

`-c:v libx264` chooses the x264 video codec.

`-c:a aac` chooses the audio codec.

`-b:a 128k` sets the audio bitrate.

`-ac 2` sets the audio channels—choose 1 for monaural, 2 for stereo.

`-ar 48000` sets the audio sample rate.

`TOS_1080p.mp4` sets the output file name.

If you want to keep the same channels and sample rate as the original file you wouldn't need to include those parameters in the command line argument.

OK, now you have the technical background to build an encoding ladder using two-pass encoding. You'll learn that in the next chapter.

Chapter 10: Multipass Encoding

	Width	Height	Frame Rate	Video Bitrate	Peak Bitrate	Buffer	Profile	Entropy	Key-frame	B-frame	Ref frame	Audio Bitrate
234p	416	234	15	145,000	159,500	145,000	Baseline	CAVLC	2	NA	10	64,000
270p	480	270	30	365,000	401,500	365,000	Baseline	CAVLC	2	NA	10	64,000
360p_l	640	360	30	700,000	770,000	700,000	Baseline	CAVLC	2	NA	10	64,000
360p_h	640	360	30	1,200,000	1,320,000	1,200,000	Main	CAVLC	2	3	10	96,000
720p_l	1,280	720	30	2,400,000	2,640,000	2,400,000	High	CABAC	2	3	10	128,000
720p_h	1,280	720	30	3,100,000	3,410,000	3,100,000	High	CABAC	2	3	10	128,000
1080p	1,920	1,080	30	5,200,000	5,720,000	5,200,000	High	CABAC	2	3	10	128,000

Table 10-1. Our valedictory exercise.

Chapter by chapter, we've learned various aspects of encoding; in this chapter we'll pull them all together, encoding the ladder shown in Table 10-1. Specifically, in this chapter, you will learn:

- additional rules for using two-pass encoding with FFmpeg

- how to produce audio-only and video-only files with FFmpeg

- how to produce multiple files most efficiently with FFmpeg.

Multiple-File Encoding in FFmpeg

Back in Chapter 4, you learned about two-pass encoding with FFmpeg, so let's start there. The text below shows a simple two-pass argument, with explanations for the syntax

```
ffmpeg -y -i TOS_1080p.mov -c:v libx264 -b:v 5200k -pass 1 -f mp4 NUL &&

ffmpeg -i TOS_1080p.mov -c:v libx264 -b:v 5200k -pass 2 TOS_1080p.mp4
```

Batch 10-1. A simple two-pass encoding exercise.

Here's an explanation of the commands used.

Line 1:

-y overwrites previous log file.

-i identifies the input file.

-c:v libx264 sets the video codec to x264.

-b:v 5200k sets the bitrate at 5200 kbps.

`-pass 1` identifies string as the first pass.

`-f mp4` chooses the MP4 output format.

`NUL` creates the log file.

`&& \` tells FFmpeg to run second pass if the first pass is successful.

Line 2:

`-i` identifies the input file.

`-c:v libx264` sets the video codec to x264.

`-b:v 5200k` sets the bitrate at 5200 kbps.

`-pass 2` identifies string as the second pass.

`TOS_1080p.mp4` sets the output file name.

As we've discussed, during the first pass, the encoder gathers information about the complexity of the file, which it uses to control the data rate during the second pass. You must include all the options in the first pass in the second, although you can add more options during the second pass that aren't in the first pass.

While the first pass is usually faster than the second pass, it can be time-consuming. If you're encoding multiple files, using the same first pass for multiple outputs can save lots of time. This leads to three questions.

1. When can you use the first pass more than once?

When building your encoding ladder, you'll change resolution, video bitrate, and maybe H.264 profile, keyframe interval, and audio values. You can use the same first pass argument when changing video resolution and all video bitrate values, but not frame rate, keyframe interval, or H.264 profile. If you use different settings for these options in the second pass than in the first, you'll see the error message shown in Figure 10-1.

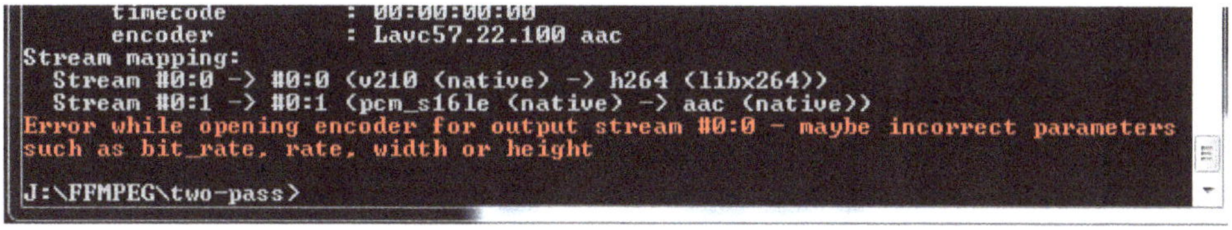

Figure 10-1. This is what you see if your second pass conflicts with your first pass.

Accordingly, for the seven files in the encoding ladder in Figure 10-1, you would need four first passes, as follows.

- **First Pass 1** 1080p, 720p_h, 720p_l

- **Second Pass 1** 360p_h (different profile from first three)

- **Third Pass 1** 360p_l and 270p (different profile)

- **Fourth Pass 1** 232p (different frame rate)

Since the files with the longest encoding time will be the top three, you'll get the most benefit from reusing the first pass there.

Obviously, if you use the High profile for all streams, as Apple recommends, you can use the same first pass for all files except for the lowest-quality stream with the different frame rate. Going beyond the configurations typically adjusted in an ABR group, note that you also can't use the same first pass when changing B-frame or reference frame values in the second pass.

2. Which parameters need to be in the first and second pass?

I'm sure someone knows, but I don't. I just know what has worked and hasn't worked.

Typically, I include x264 preset; target bitrate (but not maximum or bufsize); and B-frame, reference frame, and keyframe settings. I do not typically include video resolution or audio settings, although some sources say audio settings are essential so I'm including them in the command lines that follow. I do not include H.264 profile unless I'm encoding to the Main or Baseline profile; it has not been necessary for encoding to the High profile. That is, if your output file will be Main or Baseline, you need to include that in both the first and second passes. If it's the High profile, you can leave it out of both.

3. Which set of parameters do you include in the first pass?

We're encoding three files with three different resolutions and data rates with a single first pass—which configuration do you use for the first pass? The knee-jerk reaction is to use the highest-quality pass, although most experts recommend a stream somewhere in the middle of the ladder. I created a simple encoding ladder to test this theory, and encoded three ways, as shown in Figure 10-2. The first used the 1080p parameters in the first pass, the second the 720p parameters, and the third the 360p parameters. The encoding parameters in the second three passes were identical in all test cases.

Pass 1: 1080p params	Pass 1: 720p params	Pass 1: 360p params
Pass 2: 1080p	Pass 2: 1080p	Pass 2: 1080p
Pass 2: 720p	Pass 2: 720p	Pass 2: 720p
Pass 2: 360p	Pass 2: 360p	Pass 2: 360p

Figure 10-2. Identifying the optimal encoding parameters for the first pass.

Which configuration delivered the best quality? Table 10-2 tells the tale, and shows that on average, the 360p first pass configuration delivered the best overall quality. Note that the overall quality delta is very small, and it's only one test, so if I were starting a massive transcode of my content, I'd probably test with a few more streams before finalizing my strategy.

TOS	1080p First Pass	720p First Pass	360p First Pass	Delta
1080p	34.99	35.14	35.09	0.41%
720p	33.36	33.24	33.46	0.65%
360p	32.93	33.00	32.97	0.20%
Average	33.76	33.79	33.84	0.42%

Table 10-2. The 360p pass delivered the best quality by a hair.

In the ABR group shown in Table 10-2, I would use the parameters in the 720p_h file for the first pass for the three top-quality files; then the 270p configuration for the 270p and 360p files.

Extracting Audio or Video

Sometimes you'll need to create an audio-only or video-only stream. For example, you'll need an audio-only stream for DASH and perhaps HLS. You can produce an audio-only stream via the –vn argument. I'll duplicate the same first pass argument from above for simplicity, and add the second pass.

```
ffmpeg -y -i TOS_1080p.mov -c:v libx264 -b:v 5200k -pass 1 -f mp4 NUL && \

ffmpeg -i TOS_1080p.mov -vn -c:a aac -b:a 128k -ac 2 -ar 48000 -pass 2
no_vid.mp4
```

Batch 10-2. Two-pass encoding for the audio only file.

In the second pass, the –vn says skip the video (video? no!), while the other parameters identify the audio codec and set encoding parameters. Note that the audio-only file can use either the .mp4 or .aac extension.

What about excluding audio and producing a video only .mp4 file? Here you would add the –an argument (audio? no!) to produce the video-only .mp4 file.

```
ffmpeg -y -i TOS_1080p.mov -c:v libx264 -b:v 5200k -pass 1 -f mp4 NUL && \

ffmpeg -y -i TOS_1080p.mov -c:v libx264 -an -b:v 5200k -pass 2 no_audio.mp4
```

Batch 10-3. Two-pass encoding for the video only file.

You could also produce this file as a video-only .h264 file by specifying that extension.

> **Tip:** *Note that you can extract audio or video in a single pass argument as well. I include this discussion in this chapter because you're most likely to produce audio-only or video-only files when producing ABR files. That's the theory, anyway.*

Putting it All Together

Here's the command line argument for the first three files and an audio-only file.

```
ffmpeg -y -i TOS_1080p.mov -c:v libx264 -s 1280x720 -preset medium -g 48
-keyint_min 48 -sc_threshold 0 -bf 3 -b_strategy 2 -refs 5 -b:v 3100k
-c:a aac -b:a 128k -ac 2 -ar 48000 -pass 1 -f mp4 NUL && \

ffmpeg -i TOS_1080p.mov -c:v libx264 -preset medium -g 48 -keyint_min 48
-sc_threshold 0 -bf 3 -b_strategy 2 -refs 5 -b:v 5200k -maxrate 5720k
-bufsize 5200k -c:a aac -b:a 128k -ac 2 -ar 48000 -pass 2
TOS_1080p.mp4

ffmpeg -i TOS_1080p.mov. -c:v libx264 -s 1280x720 -preset medium -g 48
-keyint_min 48 -sc_threshold 0 -bf 3 -b_strategy 2 -refs 5 -b:v 3100k
-maxrate 3410k -bufsize 3100k -c:a aac -b:a 128k -ac 2 -ar 48000 -pass 2
TOS_720p_h.mp4

ffmpeg -i TOS_1080p.mov -c:v libx264 -s 1280x720 -preset medium -g 48
-keyint_min 48 -sc_threshold 0 -bf 3 -b_strategy 2 -refs 5 -b:v 2400k
-maxrate 2640k -bufsize 2400k -c:a aac -b:a 128k -ac 2 -ar 48000 -pass 2
TOS_720p_l.mp4

ffmpeg -i TOS_1080p.mov -vn -c:a aac -b:a 128k -ac 2 -ar 48000 -pass 2 TOS_
audio.mp4
```

Batch 10-4. Batch file to produce the first three files in the encoding ladder, plus an audio file.

As discussed elsewhere, there are a lot of parameters that you probably don't need in Batch 10-4. For example, if the input audio is stereo 48 kHz, and you want that in the output file, you could omit those parameters. You don't need to include your own B-frame or reference frame settings, or even specify the Medium preset, since that's the default. On the other hand, the settings you absolutely need to input are codec, resolution (if changed from input), keyframe interval, data rate control and audio bitrate.

That's it for multipass MP4 encoding. In the next chapter, we'll build on this and use FFmpeg to encode for HTTP Live Streaming.

Chapter 11: Producing HLS and DASH

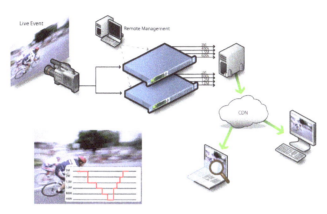

Figure 11-1. Overview of adaptive streaming. From an image by Inlet Technologies.

In the last chapter, you learned how to efficiently encode multiple files to MP4 format with FFmpeg. To use these files for adaptive bitrate streaming, however, you'll have to take one final step, which is packaging in the appropriate ABR format. In this chapter, you'll learn how to produce output for HTTP Live Streaming and DASH. Specifically, you will learn:

- how ABR technologies work

- how to produce HTTP Live Streaming (HLS) packaging with FFmpeg

- how to create and check HLS presentations with Apple Media File Segmenter, Variant Playlist Creator, and Media Stream Validator

- how to create DASH output with MP4 Box.

How ABR Technologies Work

At a high level, adaptive streaming technologies work as shown in Figure 11-1. A live or video-on-demand (VOD) source is encoded into multiple streams. These streams are distributed to different clients based upon available bandwidth, compatibility, and playback horsepower. In the early days of adaptive streaming, a dedicated server was required to communicate with the player, change streams, and dole out the necessary bits when required. These servers cost money, which increased costs and had limited capacity, which limited audience size.

Since then, the market has transitioned to HTTP-based technologies like Apple's HTTP Live Streaming, Microsoft's Smooth Streaming, and the Dynamic Adaptive Streaming over HTTP (DASH) standard. You can deploy all these technologies from standard HTTP origin servers.

All HTTP-based technologies work similarly. During encoding, you produce media files and manifest (also called playlist) files. In Figure 11-2, the .ts MPEG-2 transport stream files are the media files, which are usually segmented into short files between two and 10 seconds long.

The .m3u8 files are the manifest files. Note that there are three streams and four manifest files in our example. The master manifest file, on the left in the figure, points to the URLs of the manifest files for each stream, which point to the URL of each chunk on the HTTP server. You link to the master manifest file on your website, and the player takes it from there.

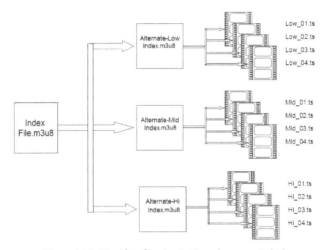

Figure 11-2. Manifest files (.m3u8) and segments (.ts).

During playback, the player retrieves the master manifest file, and then the first segment from the first stream listed in the manifest file. The player monitors a number of heuristics including buffer condition, dropped frames, and the like, and determines when a stream change, up or down, is required. To switch streams, the player checks back with the master manifest file, finds the location of the appropriate stream, and retrieves the next segment.

Since the player is in charge of all operation, you can run HTTP-based ABR technologies on standard web servers, eliminating both the cost and the capacity issues of dedicated streaming servers. In addition, because they distribute standard HTTP segments, these packets can be stored in the caching mechanisms used by many organizations and content delivery networks (CDNs), reducing bandwidth costs and improving quality of service. For all these reasons, the market has almost completely transitioned to HTTP-based adaptive technologies.

From Segments to Byte-Range Requests

When HTTP-based adaptive streaming originated, each file in the encoding ladder had to be broken into separate chunks, which created an administrative and storage nightmare. Since then, all ABR technologies have incorporated the ability to retrieve segments from a properly formatted video file via byte-range requests. Instead of retrieving separate files, the player retrieves a specific section from a single file, simplifying file creation and distribution. You'll see what that looks like in a moment.

Accordingly, when creating HLS and DASH output, you can elect to produce either a single file for each stream or multiple segments for each stream; typically I recommend the former. As you'll see, this decision is controlled via a flag in the command line.

Packaging HLS Files

To review, HLS is an Apple format that's been widely adopted by other devices, browsers, and players, including Android, Microsoft Edge, and many smart TVs and OTT boxes. Packaging HLS files involves three steps—two that FFmpeg can do, one that it can't. These are:

1. Produce a .ts file or file segments. FFmpeg can do this.

2. Produce a .m3u8 playlist file for each file. FFmpeg can do this.

3. Create a master .m3u8 file for the entire presentation. FFmpeg can't do this, at least as of June 2017. Apple tool Variant Playlist Creator can, and I detail how that works below.

In this section, you'll learn how to package HLS content from existing MP4 files, which is the workflow I recommend. You'll also learn how to create HLS output from a single mezzanine file. Then, I'll describe how to create a master .m3u8 file manually and with the Apple tool.

Packaging Existing MP4 Files

Using the command line arguments shown last chapter, we created four MP4 files: the top three files in our encoding ladder (TOS_1080p.mp4, TOS_720p_h.mp4, TOS_720p_l.mp4) and an audio-only file (TOS_audio.mp4). To package these files, we need to create output in the required MPEG-2 transport stream container format, as well as the necessary manifest files. Here are the FFmpeg commands to make that happen.

```
ffmpeg -i TOS_1080p.mp4 -bsf:v h264_mp4toannexb -codec copy -f hls
-hls_time 6 -hls_list_size 0 -hls_flags single_file TOS_1080p.m3u8

ffmpeg -i TOS_720p_h.mp4 -bsf:v h264_mp4toannexb -codec copy -f hls
-hls_time 6 -hls_list_size 0 -hls_flags single_file TOS_720p_h.m3u8

ffmpeg -i TOS_720p_l.mp4 -bsf:v h264_mp4toannexb -codec copy -f hls
-hls_time 6 -hls_list_size 0 -hls_flags single_file TOS_720p_l.m3u8
```

```
ffmpeg -i TOS_audio.mp4 -codec copy -f hls -hls_time 6 -hls_list_size
0 -hls_flags single_file TOS_audio.m3u8
```

Batch 11-1. Converting existing MP4 files to HLS output.

Here's an explanation of the new commands used.

`-bsf:v h264_mp4toannexb` is necessary when converting from MP4 to .ts. Otherwise, you'll see a "bitstream malformed" error.

`-codec copy` tells FFmpeg to copy the encoded streams, rather than transcoding them.

`-f hls` sets output as HLS.

`-hls_time 6` sets a segment size of six seconds, which is the segment duration specified by Apple in the HLS Authoring Specs (bit.ly/A_Devices_Spec).

`-hls_list_size 0` ensures that FFmpeg includes all segments in the .m3u8 file. If not included, the .m3u8 file will contain only the first five segments.

`-hls_flags single_file` creates a single output file and a playlist that specifies byte-range requests rather than separate segments (Figure 11-3, on the right). If excluded, FFmpeg will create separate segments and a manifest file that points to those segments (Figure 11-3, on the left). Note that you need clients compatible with HLS version 4.0 and above to use a single file with byte-range requests.

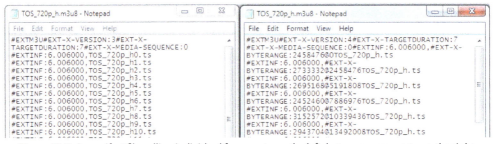

Figure 11-3. A manifest file calling individual fragments on the left, byte-range requests on the right.

Note: *There are many more HLS options, which you can find documented at* [bit.ly/FF_HLS](bit.ly/FF_HLS).

Creating HLS Output from Scratch

I don't recommend creating HLS output from scratch because you'll be unable to reuse the MP4 files that FFmpeg has to encode before packaging into the MPEG-2 transport stream, which is the time-consuming step. So, if you later decide to support DASH, you'll have to perform these lengthy transcodes again. It's better to encode to MP4 format, archive those, and then use them for DASH. If your workflow absolutely requires mezz file to HLS conversion, here's how to do it.

Basically, it's a combination of Batch 10-4 from last chapter, and Batch 11-1 from this chapter. That is, you simply add all the HLS specific arguments after `-f` in Batch 11-1 to the Batch 10-4 arguments, and change the output file extension. Here are the arguments to create the top three files in the encoding ladder, as well as the audio file.

```
ffmpeg -y -i TOS_1080p.mov -c:v libx264 -s 1280x720 -g 48 -keyint_min
48 -sc_threshold 0 -bf 3 -b_strategy 2 -refs 10 -b:v 3100k -c:a aac
-b:a 128k -ac 2 -ar 48000 -pass 1 -f HLS -hls_time 6 -hls_list_size 0
-hls_flags single_file NUL && \

ffmpeg -i TOS_1080p.mov -c:v libx264 -g 48 -keyint_min 48 -sc_
threshold 0 -bf 3 -b_strategy 2 -refs 10 -b:v 5200k -maxrate 5720k
-bufsize 5200k -c:a aac -b:a 128k -ac 2 -ar 48000 -pass 2 -f hls -hls_
time 6 -hls_list_size 0 -hls_flags single_file TOS_1080p.m3u8

ffmpeg -i TOS_1080p.mov -c:v libx264 -s 1280x720 -g 48 -keyint_min 48
-sc_threshold 0 -bf 3 -b_strategy 2 -refs 10 -b:v 3100k -maxrate 3410k
-bufsize 3100k -c:a aac -b:a 128k -ac 2 -ar 48000 -pass 2 -f hls
-hls_time 6 -hls_list_size 0 -hls_flags single_file TOS_720p_h.m3u8

ffmpeg -i TOS_1080p.mov -c:v libx264 -s 1280x720 -g 48 -keyint_min 48
-sc_threshold 0 -bf 3 -b_strategy 2 -refs 10 -b:v 2400k -maxrate
2640k -bufsize 2400k -c:a aac -b:a 128k -ac 2 -ar 48000 -pass 2 -f
hls -hls_time 6 -hls_list_size 0 -hls_flags single_file TOS_720p_l.m3u8

ffmpeg -i TOS_1080p.mov -vn -c:a aac -b:a 128k -ac 2 -ar 48000 -pass
2 -f hls -hls_time 6 -hls_list_size 0 -hls_flags single_file TOS_audio.
m3u8
```

Batch 11-2. Converting an MOV file to HLS output.

You can refer to the sections earlier for descriptions of the various commands.

Creating the Master Playlist File

HLS requires two types of playlist files: a media playlist for each stream of content, and a master playlist that links to all the media playlist files. The media playlist points to the segments or byte-range requests in the file (Figure 11-3) and that's the .m3u8 file we created with FFmpeg earlier. The master manifest is what we're examining now, and it has several jobs.

- First, it lists the encoding characteristics of the available streams so the player can identify which it can play and which it can't.

- Second, it identifies the location of the playlists for those streams so the player knows where to find the streams.

- Finally, it tells the player which layer to retrieve and start playing first.

Figure 11-4 shows a master .m3u8 manifest file produced by desktop encoder Sorenson Squeeze, along with a screenshot from FileZilla (a free FTP utility) showing the files uploaded to a website. If you navigate to bit.ly/test_hls in Safari on the Mac or in Microsoft Edge, you can play the HLS files uploaded there. Note that this won't work in Chrome or Firefox since neither supports HLS natively.

As highlighted in the figure, there are folders for each stream with a master playlist (ZOOLANDER_1080p.m3u8) at the root. Each folder contains the media files for that stream and a media playlist file like that shown on the left in Figure 11-3. The master playlist points to the media playlist files (called index.m3u8) in the individual folders Sorenson Squeeze output in the structure shown. As long as you upload the files to your web server without changing the structure or the folder or file names, it should work just fine.

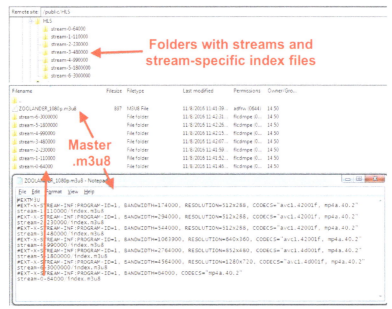

Figure 11-4. A master .m3u8 file and files produced by Sorenson Squeeze.

If you produce segments as separate files, you should strongly consider the folder-based structure shown. However, if you produce a single file with segments retrieved via byte-range requests, you can place all content and manifest files in the same folder.

The master .m3u8 file is open in Notepad on the bottom of the figure. The structure of each #EXT-X-STREAM-INF: entry is <attribute list><URI>, or Uniform Resource Identifier, which is the location of the index file for that stream. For example, on the bottom of Figure 11-4, in the last line of Notepad, you see a 64,000 bps audio-only stream where the bottom line is stream-0-64000/index.m3u8. As the arrow shows, this points to that folder on the web server. It's basically saying, "Hey, if you want to play this stream, grab this index file."

At the start of playback, the player scans through this master playlist from the top down to identify the first compatible stream. It then finds the location of the stream-specific media playlist and downloads it. Then, it finds the location of the first available media file (or byte-range), which it retrieves and starts playing. When it's time to switch streams up or down, the player checks back with the master .m3u8 to identify other available streams, and their location.

What information must be in the master .m3u8 files? This is dictated by the HLS Specification that Apple submitted to the Internet Engineering Task Force (IETF) back in 2009, which is now up to version 20 (bit.ly/HLS_spec). If you scroll to page 27 of the spec, you'll see that there are multiple attributes that can be listed in the .m3u8 file. The spec distinguishes which descriptors must be included, which should be included, and which are optional. Table 11-1 contains a summary of the most common attributes, plus their status and description.

Variant	Status	Description
BANDWIDTH	Required	Peak bitrate in bits per second
AVERAGE-BANDWIDTH	Optional	Average bitrate in bits per second
CODECS	Should Include	Codecs in the stream (see Table 15-4)
RESOLUTION	Recommended for VIdeos	Pixel resolution
FRAME-RATE	Recommended for VIdeos	Maximum frame rate rounded to 3 decimal places. Should be included for videos with frame rates in excess of 30 fps

Table 11-1. The most common variants in HLS streams.

From my perspective, you should include the first four in every master .m3u8 playlist and FRAME-RATE when your frame rates exceed 30 fps. To assist your efforts with codecs, Table 11-2 contains a list of the audio/video codec codes, which is from Apple's "HTTP Live Streaming Overview" (bit.ly/HLS_oview).

If you're creating your own master .m3u8 file, the Apple overview answers questions like how to specify an audio-only stream, or how to add a still image to an audio stream. Check with the IETF spec for details on other optional components—like High-bandwidth Digital Content Protection (HDCP) level, subtitles, and closed captions. For basic operation, you should be able to build your own .m3u8 file by duplicating the structure and attribute list in the Squeeze file shown in Figure 11-3, and substituting in your own attribute list and location information. Or, if you have access to a Mac, you can build it with Apple's Variant Playlist Creator, which you'll learn how to do in the next section.

AAC-LC	`"mp4a.40.2"`
HE-AAC	`"mp4a.40.5"`
MP3	`"mp4a.40.34"`
H.264 Baseline Profile level 3.0	`"avc1.42001e"` or `"avc1.66.30"` Note: Use `"avc1.66.30"` for compatibility with iOS versions 3.0 to 3.1.2.
H.264 Baseline Profile level 3.1	`"avc1.42001f"`
H.264 Main Profile level 3.0	`"avc1.4d001e"` or `"avc1.77.30"` Note: Use `"avc1.77.30"` for compatibility with iOS versions 3.0 to 3.12.
H.264 Main Profile level 3.1	`"avc1.4d001f"`
H.264 Main Profile level 4.0	`"avc1.4d0028"`
H.264 High Profile level 3.1	`"avc1.64001f"`
H.264 High Profile level 4.0	`"avc1.640028"`
H.264 High Profile level 4.1	`"avc1.640029"`

Table 11-2. Audio/video codec codes for your master .m3u8 file.

Working with Apple Tools

Apple has three primary HLS-related command-line tools, which are:

- ***Media File Segmenter.*** Can segment an existing H.264 encoded MP4/MOV file and create the media playlist.

- ***Variant Playlist Creator.*** Can create a master playlist file if you've created the media playlists with mediafilesegmenter.

- ***Media Stream Validator.*** Can verify that all the links that you've uploaded to the web are working, and that the file segments all meet the recommended specifications.

All three tools are available for free at developer.apple.com/streaming under "Downloads," once you sign in with your developer ID. Once you install the utilities using the Apple-supplied installation routine, they should be available from any folder on the Mac. Let's take a brief look at all three tools.

Media File Segmenter

Again, the inputs for Media File Segmenter are media files—specifically H.264 encoded files in either .mov or .mp4 format, as well as audio files encoded as .aac or even Dolby encoded audio.

The operational syntax is simple and shown below.

```
#! /bin/bash
mediafilesegmenter -I -t6 -f /Users/janozer/TOS/360p /Users/janozer/
TOS/TOS_360p.mp4
mediafilesegmenter -I -t6 -f /Users/janozer/TOS/720p /Users/janozer/
TOS/TOS_720p.mp4
mediafilesegmenter -I -t6 -f /Users/janozer/TOS/1080p /Users/janozer/
TOS/TOS_1080p.mp4
mediafilesegmenter -I -t6 -f /Users/janozer/TOS/audio /Users/janozer/
TOS/TOS_audio.mp4
```

Batch 11-3. Segmenting files and creating media playlists in Media File Segmenter.

Here's an explanation of the command structure in the batch files I created. Note that there are many more switches available, particularly for encryption and related functions.

`mediafilesegmenter` calls the application.

   `-I` creates the property list file (.plist) which identifies important file characteristics.

   `-t6` sets the segment duration at six seconds.

   `-f /Users/janozer/TOS/360p` sets the output location. Note that you have to create these folders beforehand; Media File Segmenter won't create these folders for you.

   `/Users/janozer/TOS/TOS_360p.mp4` identifies the input file.

I'm not expert in Mac batch file creation, but to make the utility work I had to include the full path to both the target folder and source file. Figure 11-5 shows the files created after running the batch file.

Here's a brief explanation of the various outputs. You know what the media segments and media playlists are. The I-frame playlist is a playlist used to add Trick Play, fast forward, and rewind to the HLS presentation, which is beyond the scope of what I'll cover here.

Note the four .plist files, one for each media file processed. This file contains basic information about the processed file including resolution, bandwidth, average bandwidth, codecs used, and the like. If you create your master playlist using Variant Playlist Creator, you'll input the .plist file and the utility will create a wholly formed master playlist. If you create your playlists manually, you can get all necessary information about each media playlist from the .plist file.

Figure 11-5. Files created by Media File Segmenter.

Variant Playlist Creator

Variant Playlist Creator inputs the media playlist and .plist files, and outputs the master playlist. The syntax is as follows:

```
variantplaylistcreator -o master.m3u8
/Users/janozer/TOS/720p/prog_index.m3u8 /Users/janozer/TOS/TOS_720p.
plist
/Users/janozer/TOS/1080p/prog_index.m3u8 /Users/janozer/TOS/
TOS_1080p.plist
/Users/janozer/TOS/360p/prog_index.m3u8  /Users/janozer/TOS/TOS_360p.
plist
/Users/janozer/TOS/audio/prog_index.m3u8 /Users/janozer/TOS/TOS_
audio.plist
```

Batch 11-4. Creating the master playlist in Variant Playlist Creator.

Here's an explanation of the command structure.

`variantplaylistcreator` runs the utility.

`-o master.m3u8` identifies the output file name.

`/Users/janozer/TOS/720p/prog_index.m3u8` identifies the media playlist URL. Repeat for each stream included in master.

`/Users/janozer/TOS/TOS_720p.plist` identifies the .plist file URL. Repeat for each stream included in master.

Then, you repeat the structure for each of the media playlists included. You can also include `-iframe-url` to identify the URL of the I-frame playlist, if desired.

The command line includes the 720p file first, since this is the stream that most viewers should be able to maintain (see the following section), then 1080p, then 360p, then audio only. This created the master playlist shown in Figure 11-6. Note that I manually removed the references to `/Users/janozer` in the master playlist to match the navigation the presentation would have once I uploaded it to the web. I also removed any mention of closed captions.

```
master.m3u8

#EXTM3U
#EXT-X-STREAM-INF:AVERAGE-BANDWIDTH=1659547,BANDWIDTH=1931856,CODECS="mp4a.40.2,
avc1.64001f",RESOLUTION=1280x720,FRAME-RATE=23.976
720p/prog_index.m3u8

#EXT-X-STREAM-INF:AVERAGE-BANDWIDTH=3180204,BANDWIDTH=3586075,CODECS="mp4a.40.2,
avc1.640028",RESOLUTION=1920x1080,FRAME-RATE=23.976
1080p/prog_index.m3u8

#EXT-X-STREAM-INF:AVERAGE-BANDWIDTH=1160684,BANDWIDTH=1363470,CODECS="mp4a.40.2,
avc1.64001e",RESOLUTION=640x360,FRAME-RATE=23.976
360p/prog_index.m3u8

#EXT-X-STREAM-INF:AVERAGE-BANDWIDTH=132693,BANDWIDTH=134396,CODECS="mp4a.40.2"
audio/prog_index.m3u8
```

Figure 11-6. The master playlist.

Media Stream Validator

This utility checks all the links in the presentation and then certain file details you can read about in Apple Technical Note TN2235: Media Stream Validator Tool Results Explained ([bit.ly/tn2235](bit.ly/tn2235)). Running the program is simple, although it only works on files already uploaded to the web. Here's the command syntax:

```
mediastreamvalidator http://www.doceopub.com/TOS/master.m3u8
```

Batch 11-5. Checking the HLS package in Media Stream Validator.

Yup, utility name and then the URL of the master playlist file. You can see the preliminary results in Figure 11-7, which shows that all the files are present and accounted for.

```
Jans-Mac-Pro:TOS janozer$ mediastreamvalidator http://www.doceopub.com/TOS/master.m3u8
mediastreamvalidator: Version 1.2(160525)

[/TOS/master.m3u8] Started root playlist download
[audio/prog_index.m3u8] Started media playlist download
[720p/prog_index.m3u8] Started media playlist download
[1080p/prog_index.m3u8] Started media playlist download
[360p/prog_index.m3u8] Started media playlist download
[audio/prog_index.m3u8] All media files delivered and have end tag, stopping
[720p/prog_index.m3u8] All media files delivered and have end tag, stopping
[360p/prog_index.m3u8] All media files delivered and have end tag, stopping
[1080p/prog_index.m3u8] All media files delivered and have end tag, stopping
```

Figure 11-7. All the files are present and accounted for—a good start.

Beyond this, Media Stream Validator presents a range of other information that's particularly important when you're seeking App Store approval for an app that will play your videos. You should run this utility for all HLS presentations you upload, particularly when you're first starting out.

OK, now that we know how to produce the master playlist file, let's take a look at which file you should place first.

Which Stream First?

As mentioned earlier, the stream listed first in the .m3u8 file is the stream played first by the HLS player. Which stream should you deploy first? In Tech Note TN2224, Apple says, "Therefore, the first bit rate in the playlist should be the one that most clients can sustain." In the Apple Devices spec, Apple gets even more directive, saying, "For WiFi delivery, the default video variant(s) SHOULD be the 2000 kb/s variant. For cellular delivery, the default video variant(s) SHOULD be the 730 kb/s variant." (bit.ly/A_Devices_Spec).

Packaging for DASH with MP4Box

Packaging DASH output involves the same basic steps as HLS, although FFmpeg can't perform any of them. Instead, I'll demonstrate with an open-source tool called MP4Box.

MP4Box is one of the multiple open-source utilities created by GPAC, which includes the Osmo4 player, MP4Box, and multiple server tools. You can download the complete set of utilities at bit.ly/dlMP4box, and click through the site to read more about the tools and organization.

To package your files with MP4Box, you need to ensure that your keyframe interval divides evenly into your segment size, which I'll show you how to set below. I'll be working with the same four output files creating earlier in the section titled "Putting It All Together," which I encoded using a keyframe interval of 48 frames, or two seconds.

As mentioned earlier, MP4Box can only perform this DASHing process with MP4 inputs (hence the name), which is why I produced the audio/video and audio-only files in MP4 format. Note that most DASH players want separate audio video streams, rather than interleaved or muxed audio/video streams as with HLS and standard MP4 files. Rather than create a set of video only MP4 files, I'll use the `#video` option in the command line to tell MP4Box to only insert video from the MP4 files into the MPD file. Here's the command line.

```
mp4box.exe -dash 4000 -rap -dash-profile dashavc264:onDemand
-bs-switching no TOS_720p_l.mp4#video TOS_720p_h.mp4#video
TOS_1080p.mp4#video TOS_audio.mp4
```

Batch 11-6. Packaging to DASH with MP4Box.

Here's an explanation of the switches shown.

`mp4box.exe` runs the program.

`-dash 4000` sets the segment length (in milliseconds), so this is four seconds.

`-rap` forces segments to begin at random access points.

`-dash-profile dashavc264:onDemand` makes sure a single file accessed via byte-range requests is output, rather than separate segments.

`TOS_720p_l.mp4#video TOS_720p_h.mp4#video TOS_1080p.mp4#video TOS_audio.mp4` defines the input files, using the #video to insert only the video into the fragmented MP4 file.

The output from this function are the files shown in Figure 11-8, which you upload to your web server to make available to your viewers, linking to the .mpd file. The four .mp4 files are each elementary streams, meaning that they contain only audio or video, but not both. The .mpd file directs the player which streams to retrieve and how to mux them for playback. In this case, the player is retrieving a byte-range request within each file, since we chose this approach rather than producing separate segments.

Name	Date modified	Type	Size
TOS_1080p_dash.mpd	11/8/2016 4:53 PM	MPD File	3 KB
TOS_720p_h_track1_dashinit.mp4	11/8/2016 4:53 PM	VLC media file (.m...	51,267 KB
TOS_720p_l_track1_dashinit.mp4	11/8/2016 4:53 PM	VLC media file (.m...	29,427 KB
TOS_1080p_track1_dashinit.mp4	11/8/2016 4:53 PM	VLC media file (.m...	73,087 KB
TOS_audio_track1_dashinit.mp4	11/8/2016 4:53 PM	VLC media file (.m...	1,932 KB

Figure 11-8. The output from DASHing the input files.

MP4Box contains many more useful options, which you can read about at bit.ly/mp4boxdash. You can read up on DASH in general at bit.ly/mp4boxdash2. French company Streamroot has done a nice job posting several articles on encoding and packaging for DASH, which you can read at bit.ly/srdashing. I tried to accurately incorporate Streamroot's direction into this chapter, but if you're having problems or are looking for more advanced tutorials, go there.

Tip: *MP4Box can also encode the files for you if that's your preference, though I haven't demonstrated how in this book.*

So, we're good for encoding and packaging your ABR streams. In the next chapter, you'll learn how to produce HEVC output.

Chapter 12: Encoding HEVC

HEVC is the standards-based successor to H.264 that's primarily targeted towards Smart TVs and set-top boxes. As with H.264, there are multiple HEVC codecs out there; the codec that's available in FFmpeg is called x265. As you'll see, working with x265 is very similar to working with x264 except that the encoding times are much longer and the quality is much better.

In this chapter, you'll learn:

- considerations for encoding HEVC, including a look at HEVC encoding profiles
- how to encode using the x265 codec in FFmpeg.

During the chapter, I'll identify FFmpeg commands for various x265 parameters. However, you'll have to apply them in a special way, which I detail towards the end of the chapter.

What is HEVC

HEVC is a standards-based codec created by the same groups that created H.264. Like H.264, you produce HEVC by choosing different profiles, and x265 offers presets and tuning mechanisms.

You can produce x265 in FFmpeg, which I'll demonstrate in this chapter, or with the x265 executable that you can download from http://x265.org/. The quality should be identical, but you'll have to convert your source files to YUV or Y4M formats to input them into x265, which is time-consuming and can consume tons of disk space. With FFmpeg, you can input the same files you've been using for H.264, which is faster and easier.

Like H.264, HEVC is a codec, not a container format. While you can package HEVC in multiple container formats, single files are typically encoded in the MP4 container format, while adaptive bitrate files are typically packaged in Dynamic Adaptive Streaming over HTTP (DASH) format.

Basic HEVC Encoding Parameters

With this as background, let's talk basics of HEVC encoding. At a high level, all non-H.264 specific lessons learned in previous chapters regarding bitrate control, I-, B-, and P-frames, resolution, and frame rate apply here. Beyond these, you'll have to choose a profile, a preset, and if desired, a tuning mechanism. Of course, you'll have to choose the codec first, which you do with the following command string.

```
-c:v libx265
```

Since AAC is the audio codec for HEVC, no changes are necessary there.

HEVC Profiles

Most HEVC encoding tools let you select the Main or Main 10 profile. The Main profile supports 8 bits per sample, which allows 256 shades per primary color, or 16.7 million colors in the video. In contrast, the Main 10 profile supports up to 10 bits per sample, which allows up to 1024 shades and over 1 billion colors. Of course, you'll need a 10-bit display to see the extra colors, which most potential viewers don't have at this point. That's because the real targets of Main 10 output are high-dynamic-range (HDR) displays, which are just starting to ship in quantity in 2017.

In addition, if your video has an 8-bit color depth, which most formats do, encoding in 10-bit won't add the colors and improve video quality. On the other hand, some experts argue that processing in 10-bit color may improve the encoding precision of 8-bit source videos, even if it doesn't add colors. My tests didn't quite confirm this claim, as you can see in Table 12-1.

720p - x265	Main	Main 10	Delta
Tears of Steel	37.05	37.73	1.84%
SIntel	41.37	41.25	-0.29%
Big Buck Bunny	37.21	37.16	-0.13%
Talking Head	41.15	41.15	0.00%
Freedom	39.70	39.57	-0.31%
Haunted	39.56	41.78	5.61%
Average	**39.34**	**39.77**	**1.12%**

Table 12-1. Main 10 delivered slightly higher quality than Main.

Here, I encoded 8-bit source videos using the Main and Main 10 profiles and otherwise identical features. Although the Main 10 encoded videos averaged slightly higher quality, four of the six videos were either about the same or worse quality.

Before chasing the extra quality some claim Main 10 can deliver, note that if you encode your video using the Main 10 profile, only Main 10 compatible decoders can play the video. Most early HEVC players were not Main 10 compatible, so if you're distributing HEVC videos to the general public, rather than to specific smart TVs or set-top boxes, there is a compatibility risk (see Figure 12-1).

So, I recommend using the Main profile for general-purpose distribution, even if your video is 10-bit in origin. If you're distributing to known Main 10 compatible HEVC decoders, you should consider encoding with Main 10 even if your video is 8-bit in origin. Obviously, if you're encoding HDR video, which is beyond the scope of this tutorial, you'll need to use Main 10.

Figure 12-1. Only Main 10 compatible players can play Main 10 files.

Note: *To encode to Main 10 using FFmpeg and x265, you'll have to download or compile a Main 10-specific version. You can't call the Main 10 x265 codec from the standard downloadable version of FFmpeg.*

Quick Summary: HEVC Profiles

1. If you're encoding video for general-purpose distribution, use the Main profile for the broadest possible compatibility.

2. If you're producing for a platform or platforms with known Main 10 compatibility, encode using the Main 10 profile, whether the source footage is 8-bit or 10-bit.

As mentioned, to create Main 10 output with FFmpeg, you'll need an FFmpeg build with 10-bit libx265. Once you have that, you can specify the profile using one of these strings:

```
-profile main
```

```
-profile main10
```

x265 Presets

The x265 encoding presets share the same names as the x264 presets, and Table 12-2 shows their comparative quality. As you can see, the Ultrafast preset always produced the lowest quality, and Placebo the highest. Interestingly, quality actually dropped after the Superfast preset, and didn't surpass that level until the Fast preset. Overall, the average difference between the highest and lowest scores was 6.7 percent.

	Ultrafast	Superfast	Veryfast	Faster	Fast	Medium	Slow	Slower	Veryslow	Placebo	Total Delta
Tears of Steel	37.25	38.06	38.04	38.05	38.34	38.39	38.84	38.86	38.93	39.00	4.70%
Sintel	35.87	36.89	36.66	36.67	37.11	37.25	37.74	37.79	37.90	37.97	5.86%
Big Buck Bunny	36.10	37.65	37.61	37.60	37.91	38.26	38.70	38.89	39.03	39.18	8.54%
Freedom	38.16	39.01	38.45	38.46	38.71	38.98	39.36	39.44	39.52	39.58	3.72%
Haunted	41.36	41.77	41.39	41.39	41.55	41.68	41.97	41.92	41.97	42.02	1.60%
Screencam	44.03	46.70	46.55	46.54	46.78	47.12	48.31	48.69	48.99	49.34	12.07%
Tutorial	42.46	47.14	46.46	46.42	46.52	47.19	48.35	47.65	48.02	48.53	14.31%
Average	38.64	39.51	39.30	39.31	39.58	39.74	40.13	40.18	40.27	40.35	6.70%

Table 12-2. PSNR quality by video file and encoding preset.

How does encoding time factor in? This is shown in Figure 12-2. Here I've normalized encoding time on a scale from 0 to 100, with the time of the Ultrafast encode set to 0. The quality side shows the percentage of quality each preset delivers as compared to the Placebo preset that delivers maximum quality or 100%.

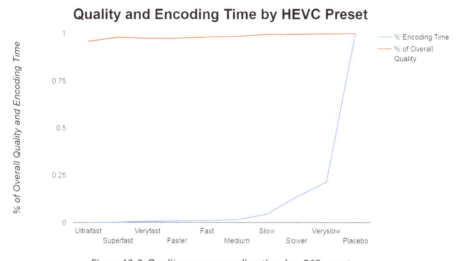

Figure 12-2. Quality versus encoding time by x265 preset.

From an encoding time perspective, the needle barely even moves until the Medium preset, and the quality jump from Medium to Slow makes the Slow preset look like an obvious move. At Slow, you're just under 99.47 percent of total quality, and encoding time for the higher quality presets really starts to jump. If you're running out of capacity, it's worth experimenting with Superfast, as the bang for your encoding time buck is substantial. To put the numbers in perspective, PSNR for Superfast averaged 39.51 dB, while Slow averaged 40.13 dB, which isn't a difference that most viewers would notice.

You choose x265 presets just like x264 presets, using this string:

```
-preset [preset]

-preset veryslow
```

As with x264, if you don't specify a preset, FFmpeg will use the default Medium preset.

> **Tip:** *Note that you can see the exact configuration options used for each preset at bit.ly/x265_pre, a page created by x265 developer MulticoreWare.*

Our HEVC Encoding Ladder

Table 12-3 shows an encoding ladder for HEVC encodes. Basically, I decreased the data rates for the H.264 encodes by about 50%, eliminated the lowest rung of the ladder and added rungs for 1440p and 2160p. While this is as generic as you can get, it gives us a starting point for the HEVC encodes that I'll demonstrate below.

	Width	Height	Frame Rate	Video Bitrate	Peak Bitrate	Buffer	Profile	Preset	Key-frame	B-frame	Ref frame	Audio Bitrate
270p	480	270	30	220,000	242,000	220,000	Main	Slow	2	3	5	64,000
360p_l	640	360	30	400,000	440,000	400,000	Main	Slow	2	3	5	64,000
360p_h	640	360	30	720,000	792,000	720,000	Main	Slow	2	3	5	96,000
720p_l	1,280	720	30	1,000,000	1,100,000	1,000,000	Main	Slow	2	3	5	128,000
720p_h	1,280	720	30	1,800,000	1,980,000	1,800,000	Main	Slow	2	3	5	128,000
1080p_l	1,920	1,080	30	2,500,000	2,750,000	2,500,000	Main	Slow	2	3	5	128,000
1080p_h	1,920	1,080	30	4,000,000	4,400,000	4,000,000	Main	Slow	2	3	5	128,000
1440p	2,560	1,440	30	6,000,000	6,600,000	6,000,000	Main	Slow	2	3	5	128,000
2160p	3,840	2,160	30	10,000,000	11,000,000	10,000,000	Main	Slow	2	3	5	128,000

Table 12-3. Encoding ladder for HEVC encodes.

Note that we don't have a lot of commercial comparisons we can use because there's very little HEVC-encoded content available for testing. Netflix reportedly encodes *House of Cards* at 16 Mbps, which blows my numbers out of the water, but you must start somewhere.

> **Tip:** *Just as I was finishing this book, I wrote an article on HDR production for Streaming Media Magazine. Look for it for a solid overview of HDR production. One fabulous article I found in my research was entitled, HDR Video Part 5: Grading, Mastering, and Delivering HDR, and it includes a detailed description of how to produce HDR with an FFmpeg front end called Hybrid. If you need to learn how to produce HDR with FFmpeg, check out the article at bit.ly/hdr_ffmpeg.*

x265 and FFmpeg

As mentioned above, encoding to x265 is easier in FFmpeg than using the x265 executable because you don't have to pre-convert the files to YUV/Y4M. On the other hand, there really is very limited documentation for the x265 controls available in FFmpeg, which is a pain.

In contrast, x265 is very well documented (see bit.ly/x265_documentation). In theory, you can add any x265 configuration option to an FFmpeg command script by adding -x265-params to the FFmpeg command string and adding the x265 parameters after that (see bit.ly/x265_ffmpeg). In practice, however, it's not quite that simple, and I had to evolve into the following approach to produce the desired format. I'm not sure that this is the only way to do it, or even the best way, but it worked for me.

1080p Conversion

Here's a script I used to create the 1080p and lower ladders from Table 12-3 from 1080p source.

```
ffmpeg -y -i TOS_1080p.mov -c:v libx265 -preset slow -x265-params
profile=main:keyint=48:min-keyint=48:scenecut=0:ref=5:bframes=3:b-
adapt=2:bitrate=4000:vbv-maxrate=4400:vbv-bufsize=4000 -an -pass 1 -f mp4
NUL && \

ffmpeg -i TOS_1080p.mov -c:v libx265 -preset slow  -x265-params
profile=main:keyint=48:min-keyint=48:scenecut=0:ref=5:bframes=3:b-
adapt=2:bitrate=4000:vbv-maxrate=4400:vbv-bufsize=4000 -an -pass 2
TOS_1080p_h.mp4

ffmpeg -i TOS_1080p.mov -c:v libx265 -preset slow  -x265-params
profile=main:keyint=48:min-keyint=48:scenecut=0:ref=5:bframes=3:b-
adapt=2:bitrate=2500:vbv-maxrate=2750:vbv-bufsize=2500 -an -pass 2
TOS_1080p_l.mp4

ffmpeg -i TOS_1080p.mov -c:v libx265 -s 1280x720 -preset slow -x265-
params profile=main:keyint=48:min-keyint=48:scenecut=0:ref=5:bframes=3
:b-adapt=2:bitrate=1800:vbv-maxrate=1980:vbv-bufsize=1800 -an -pass 2
TOS_720p_h.mp4

ffmpeg -i TOS_1080p.mov -c:v libx265 -s 1280x720 -preset slow  -x265-
params profile=main:keyint=48:min-keyint=48:scenecut=0:ref=5:bframes=3
:b-adapt=2:bitrate=1000:vbv-maxrate=1100:vbv-bufsize=1000 -an -pass 2
TOS_720p_l.mp4

ffmpeg -i TOS_1080p.mov -c:v libx265 -s 640x360 -preset slow  -x265-
params profile=main:keyint=48:min-keyint=48:scenecut=0:ref=5:bframes=
3:b-adapt=2:bitrate=720:vbv-maxrate=792:vbv-bufsize=720 -an -pass 2
TOS_360p_h.mp4

ffmpeg -i TOS_1080p.mov -c:v libx265 -s 640x360 -preset slow  -x265-
params profile=main:keyint=48:min-keyint=48:scenecut=0:ref=5:bframes=
3:b-adapt=2:bitrate=400:vbv-maxrate=440:vbv-bufsize=400 -an -pass 2
TOS_360p_l.mp4

ffmpeg -i TOS_1080p.mov -c:v libx265 -s 480x270 -preset slow -x265-
params profile=main:keyint=48:min-keyint=48:scenecut=0:ref=5:bframes=3:b-
adapt=2:bitrate=220:vbv-maxrate=242:vbv-bufsize=220 -an -pass 2 TOS_270p.
mp4

ffmpeg -i TOS_1080p.mov -vn  -c:a aac -b:a 128k  -pass 2  TOS_audio.mp4
```

Batch 12-1. Converting 1080p source to our HEVC encoding ladder 1080p rungs and below.

The funky things, of course, are that the size and preset are outside the `-x265-params` string while the other parameters follow it, and are tied together with colons. By this point, if you just follow carefully, you should have no trouble duplicating these results.

Note that I've included the -an switch (audio? no!) in the video outputs, and -vn (video? no!) in the final output for audio. Since I didn't need to change either the number of channels or sampling frequency, I didn't specify those in the audio string.

4K Scaling Exercise

We covered multiple examples of scaling back in Chapter 5. Since you're most likely to encounter these operations when working with 4K content, I wanted to test and make sure they still work when producing x265.

The following example inputs Tears of Steel at 3840x1714 resolution, with a display aspect ratio of 2.25:1 and outputs the encoding ladder shown in Table 12-3, save the high-quality versions of the 1080p, 720p, and 360p files, which I removed to save space. With the 4K and 2K files, I produced at full resolution (3840x2160, 2540x1440) with letterboxing, essentially using the command string shown in Batch 5-6. In all other resolutions, I produced at full resolution while cropping out the excess pixels as shown in Batch 5-5.

Here are the commands.

```
ffmpeg -y -i TOS_4k.mov -c:v libx265 -preset slow -x265-params
profile=main:keyint=48:min-keyint=48:scenecut=0:ref=5:bframes=3:b-
adapt=2:bitrate=4000:vbv-maxrate=4400:vbv-bufsize=4000 -an -pass 1 -f mp4
NUL && \

ffmpeg -i TOS_4k.mov -c:v libx265 -vf "scale=3840:2160:force_original_as-
pect_ratio=decrease,pad=3840:2160:(ow-iw)/2:(oh-ih)/2" -preset slow
-x265-params profile=main:keyint=48:min-keyint=48:scenecut=0:ref=5:bframes=
3:b-adapt=2:bitrate=10000:vbv-maxrate=11000:vbv-bufsize=10000 -an -pass 2
TOS_4K_lb.mp4

ffmpeg -i TOS_4k.mov -c:v libx265 -vf "scale=2560:1440:force_original_as-
pect_ratio=decrease,pad=2560:1440:(ow-iw)/2:(oh-ih)/2" -preset slow
-x265-params profile=main:keyint=48:min-keyint=48:scenecut=0:ref=5:bframe
s=3:b-adapt=2:bitrate=6000:vbv-maxrate=6600:vbv-bufsize=6000 -an -pass 2
TOS_2K_lb.mp4

ffmpeg -i TOS_4k.mov -c:v libx265 -preset slow -vf "scale=1920:1080:force_
original_aspect_ratio=increase,crop=1920:1080" -x265-params
profile=main:keyint=48:min-keyint=48:scenecut=0:ref=5:bframes=3:b-
adapt=2:bitrate=2500:vbv-maxrate=2750:vbv-bufsize=2500 -an -pass 2
TOS_1080p_l.mp4
```

```
ffmpeg -i TOS_4k.mov -c:v libx265  -preset slow -vf "scale=1280:720:force_
original_aspect_ratio=increase,crop=1280:720" -x265-params
profile=main:keyint=48:min-keyint=48:scenecut=0:ref=5:bframes=3:b-
adapt=2:bitrate=1000:vbv-maxrate=1100:vbv-bufsize=1000 -an -pass 2
TOS_720p_l.mp4

ffmpeg -i TOS_4k.mov -c:v libx265  -preset slow -vf "scale=640:360:force_
original_aspect_ratio=increase,crop=640:360" -x265-params
profile=main:keyint=48:min-keyint=48:scenecut=0:ref=5:bframes=3:b-
adapt=2:bitrate=400:vbv-maxrate=440:vbv-bufsize=400 -an -pass 2
TOS_360p_l.mp4

ffmpeg -i TOS_4k.mov -c:v libx265  -preset slow -vf "scale=480:270:force_
original_aspect_ratio=increase,crop=480:270" -x265-params
profile=main:keyint=48:min-keyint=48:scenecut=0:ref=5:bframes=3:b-
adapt=2:bitrate=220:vbv-maxrate=242:vbv-bufsize=220 -an -pass 2
TOS_270p.mp4

ffmpeg -i TOS_4k.mov -c:v libx265 -vf "scale=2560:1440:force_original_as-
pect_ratio=decrease,pad=2560:1440:(ow-iw)/2:(oh-ih)/2" -preset slow
-x265-params profile=main:keyint=48:min-keyint=48:scenecut=0:ref=5:bframe
s=3:b-adapt=2:bitrate=4000:vbv-maxrate=4400:vbv-bufsize=4000 -an -pass 2
TOS_2K_lbx.mp4

ffmpeg -i TOS_4k.mov -vn  -c:a aac -b:a 128k -pass 2  TOS_audio.mp4
```

Batch 12-2. Converting 4K source to our HEVC encoding ladder with letterboxing and cropping.

I checked the files and encoding, and the controls seemed to work as advertised. Give it a try!

So, that's HEVC. In the next chapter, you'll learn the ins and outs of encoding VP9.

Chapter 13: Encoding VP9

VP9 is an open-source codec from Google that was developed from technology acquired from On2 Technologies that Google purchased in February 2010. The first codec Google released from this acquisition was VP8, which was paired with the Vorbis audio codec in the WebM container. VP9 is the next iteration of the codec, which became available on June 17, 2013. VP9 will be the last VPx-based codec released by Google, as the company contributed all codec technology to the Alliance for Open Media in September 2015.

In this chapter, you'll learn:

- about VP9, including encoding parameters and container formats
- the common encoding parameters for VP9, including sample scripts for FFmpeg.

About VP9

VP9 is a completely different codec than H.264 and H.265, and while it shares some configurations like resolution, frame rate, I-frame setting and the like, many other parameters are completely different. In general, VP9 is very similar to HEVC in encoding quality and much more efficient than H.264.

VP9 is a codec, not a container format. Though you can package VP9 in multiple container formats, single files are typically encoded into the WebM container format. Adaptive bitrate files are typically packaged in DASH format. For example, as shown in Figure 13-1, YouTube uses DASH for VP9 files (nine lines down, DASH: yes).

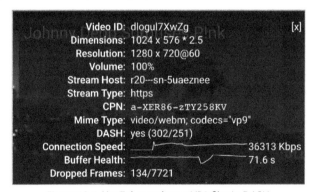

Figure 13-1. YouTube packages VP9 files in DASH.

Basic VP9 Encoding Parameters

While many encoding concepts are similar to H.264 and H.265, VP9 encoding has many unique options and Google has done a poor job documenting what many of them do and how they work. Until recently, few websites have used VP9, so application support among encoding tools is weak and each application seems to implement different features in different ways.

There's also a disconnect between the command line parameters available for Google's own encoder (vpxenc.exe) and FFmpeg's VP9-specific encoding parameters. Specifically, Google did an adequate job detailing the command line parameters for its own encoder but hasn't really documented which configurations work with FFmpeg and the appropriate command string. As with HEVC, there's a dearth of tools for analyzing your encoded files to verify that the selected encoding options are working as you want them to.

As a result, most of the sample scripts that Google has provided are basic, which seems to be the approach that most developers take. Let's examine the various options of the command string that we'll use to create our FFmpeg batch files.

Bitrate Control

According to the FFmpeg wiki (bit.ly/vp9_brcontrol), VP9 supports at least five data rate control mechanisms—including variable bitrate, constant quality, constrained quality, constant bitrate, and lossless mode. Analyzing the operation of each technique is hampered by the inability to visualize a bitstream. That said, VP9 does support VBR, and the `maxrate` control appears to limit the maximum bitrate as it does for H.264 and H.265. So, we'll use the 110% constrained VBR approach discussed throughout the book.

Note that I tried various VBV settings with some test encodes, and they had absolutely no impact on file output. Though I didn't get confirmation from Google that VBV isn't a factor in VP9 encoding, that's what my tests showed, so I'll ignore VBV for the purposes of this chapter.

Other Configuration Options

Table 13-1 shows the other options typically recommended by VP9 documentation or from other sources. The first two sets of options, VOD Recommended and Best Quality (Slowest) are from the VP9 Encoding Guide available at bit.ly/vp9_guide1. The DASH column is available in a wiki page titled "Instructions to Playback Adaptive WebM Using DASH" (bit.ly/vp9_dash). The final column are the configuration options used by JWPlayer, which shared its encoding configuration with me in early 2016 for a Streaming Media article about VP9 (bit.ly/vp9_age1).

As you'll see, I spent most of my time testing the speed option, which controls the classic encoding time/encoding quality trade-off—with 0 being the longest, highest-quality option, and 4 being the fastest, lowest-quality option. Otherwise, I pretty much went with the options JW Player used, since I knew they had been rigorously tested before implementation.

	VOD Recommended	Best Quality (slowest)	DASH	JWPlayer
Threads	8	1	default	8
Speed (1rst/2nd pass)	4/1	4/0	default	4/2
Tile-Columns	6	0	4	6
Frame Parallel	1	0	1	1
Auto-Alt-Ref	1	1	default	1
Lag-In-Frames	25	25	default	25

Table 13-1. Other VP9 configuration options.

As mentioned previously, Google does not do a great job identifying when and where you should use the various options. Here's what I could glean from different documents and encoding formulas.

- **Threads.** This allows the encoder to use multiple cores. While there's a minor quality hit, VP9 encoding is glacial without it.

- **Speed.** All encoders that specified speed used 4 for the first pass, and a lower value for the second pass. You'll see the quality/encoding time trade-off curve in a moment.

- **Tile/Columns.** With the threads command, this allows the encoder to use more than a single CPU core.

- **Frame parallel.** Here's what the VP9 Encoding Guide (bit.ly/vp9_guide) says about tile-columns and frame parallel: "Turning off tile-columns and frame-parallel should give a small bump in quality, but will most likely hamper decode performance severely."

- **-auto-alt-ref and -lag-in-frames.** These win my award for the most obtuse encoding configuration options ever (and I've seen a few). Here's the description from the VP8 Encode Parameter Guide (bit.ly/vp8_guide). "When `-auto-alt-ref` is enabled the default mode of operation is to either populate the buffer with a copy of the previous golden frame when this frame is updated, or with a copy of a frame derived from some point of time in the future (the choice is made automatically by the encoder). The `-lag-in-frames` parameter defines an upper limit on the number of frames into the future that the encoder can look.

Let's look at each encoding parameter in turn.

Threads

To test the impact of threads, I encoded twice—once with threads set at 1, once with threads set at 8. The output files were identical, with identical PSNR scores, but encoding with 8 threads cut encoding time by about 50 percent. I asked my contacts at Google about this and they replied:

Currently our multi-threaded encoder does not compromise on quality and results are identical. For multi-core machines, you should use multiple threads.

However, when we encode on the Google cloud for YouTube, accounting is often done by cores, and if you are to optimize for the encode_cores x encode_time product, then using single threads would be the best.

So, if you're running a single encode on a multiple-core computer, always set threads to 8. If you're creating your own encoder that will run multiple encodes simultaneously, you should experiment with different settings to determine which produces the best performance.

Here's the syntax for setting threads:

```
-threads 8
```

Speed

Table 13-2 shows the PSNR values for our test files (less the Tutorial file, which failed to meet encoding targets) encoded at 1080p resolution). As you can see, 0 always delivered the best quality and 4 the worst, but the average difference was only 2.54 percent. Not a huge deal. The average difference was 2.96 percent at 720p—slightly higher but still pretty minor.

Speed - 1080p	4	3	2	1	0	Delta
Tears of Steel	39.85	40.01	40.73	40.86	41.12	3.19%
Sintel	38.45	38.69	39.25	39.37	39.62	3.06%
Big Buck Bunny	38.83	39.09	39.66	39.80	39.93	2.83%
Talking Head	43.36	43.48	44.08	44.22	44.30	2.17%
Freedom	40.55	40.79	41.26	41.49	41.74	2.92%
Haunted	41.33	41.45	41.86	41.98	42.05	1.75%
Screencam	43.20	43.76	43.79	43.94	44.02	1.88%
Average	40.80	41.04	41.52	41.66	41.83	2.54%
Percentage	97.54%	98.12%	99.26%	99.61%	100.00%	

Table 13-2. Output quality by speed option.

At the lowest setting, you capture 97.54% of available quality, which advances to over 99% for both 2 and 1. Table 13-3 shows the encoding time to VP9 format at 1080p resolution.

	4	3	2	1	0
Tears of Steel	307	332	457	712	3422
Sintel	300	327	468	708	3535
Big Buck Bunny	231	257	416	604	2926
Talking Head	315	339	509	796	2703
Freedom	408	465	585	886	3796
Haunted	413	449	615	1038	4442
Screencam	120	106	218	290	1052
Average	299	325	467	719	3125

Table 13-3. Encoding time in seconds.

Figure 13-2 shows the encoding time/quality trade-off. As mentioned, once you get to the setting of 2, you're above 99% of all available quality at less than twice the encoding time of the lowest quality setting. Option 0 looks like a bad investment unless encoding time is irrelevant.

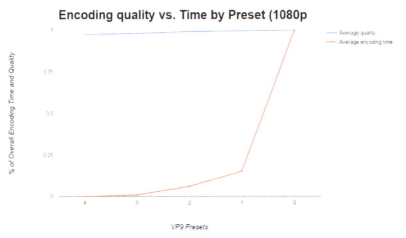

Figure 13-2. Encoding time/quality trade-off by VP9 preset.

To wrap this in a shiny bow for you, I would probably default to 2, but if I reached encoding capacity at that setting, and was forced to either buy another workstation or change to 3 or 4, I would change to 4. Even a "golden eye" viewer would have a hard time telling the difference between video with a PSNR of 40.80 compared to 41.83.

Note that most of the command strings that I reviewed used two-pass encoding, with a speed setting of 4 for the first pass, and 0 to 2 for the second. Thus, it appears pretty certain that scanning the file with a speed setting of 4 for the first pass doesn't degrade quality.

Here's how you set speed in the FFmpeg command string:

```
-speed 2
```

In a two-pass string, you would use a setting of 4 in the first pass, and 0 to 2 in the second.

Frame Parallel

I didn't experiment with this parameter because literally until the day before I shipped the book, the VP9 documents said, "Turning off tile-columns and frame-parallel should give a small bump in quality, but will most likely hamper decode performance severely." Google's corrections to its own documents just came too late for me to test this parameter.

Note that the default for frame parallel is enabled, so if you don't refer to this option in your command line script, it will be enabled. If you're the type who likes to wear a belt and suspenders (figuratively, of course), you would enable this in your command script using the following command:

```
-frame-parallel 1
```

To disable this option:

```
-frame-parallel 0
```

Tile/Columns

Tile/columns is another command that lets the encoder divide up the image and encode with multiple cores. Technically, you should customize this option by output resolution using this formula supplied by my Google contact.

> The way to figure this out is to take the width of the video, divide by 256, and see what power of 2 is at least as large as that. Mathematically, the effective parameter is:
> ```
> tile_width = floor(log2(width/256)).
> ```

So:
320x180: 0
480x270: 0
640x360: 1
848x480: 1
960x540: 1
1280x720: 2
1920x1080: 2
2560x1440: 3
3840x2160: 3
8K: 4

Saving us all some spreadsheet time, he also noted, "Yes, it does not matter if the tile columns parameter is larger than what can be supported by the format. So, using 4 will work for all."

So the bottom line is to use 4, and the encoder will step it down as needed. Given that the maximum setting appears to be 4, I have no idea why JW Player and the recommended VOD settings use 6, though it appears that you could use the US federal deficit and the encoder would step it down to 4.

Here's the FFmpeg syntax:

```
-tile-columns 4
```

-auto-alt-ref

If you have no entry for `-auto-alt-ref`, the encoder defaults to a setting of 1, which delivered slightly higher quality than a setting of 0 in my tests. For example, with the Freedom video encoded at 720p, a setting of 1 delivered a PSNR of 39.44, while a setting of 0 delivered a PSNR of 39.17, a difference of 0.69 percent. Not a huge deal by any means, but unless you know something that I don't, I would always use a setting of 1. Again, you can do this directly by including the following in your command string:

```
-auto-alt-ref 1
```

Or, you can just leave the configuration option out. If you want to disable this configuration option—and again, I'm not sure why you would—include this in your command string.

```
-auto-alt-ref 0
```

-lag-in-frames

For `-lag-in-frames`, I asked my contact, "What is the default `-lag-in-frames` value? Does this make any difference? What are the trade-offs with values here (0 to 25)? All the recommendations in the VP9 encoding guide use 25. Should I just recommend using that?"

He replied:

> `-lag-in-frames` default is 25 if the parameter is not explicitly specified. We can reduce it to up to 16 with very little change in coding efficiency. Beyond that, it starts affecting efficiency more since alt-ref frames cannot to used to their fullest potential. Note that `-lag-in-frames` needs memory to store lookahead frames. So, for 4K or 8K, one can use smaller values to prevent memory issues.

So, I recommend going with 25 for all encodes up to 4K, then switching to 16. Here's how you configure this in your command string.

```
-lag-in-frames 25
```

Advice from the Stars

While writing the Streaming Media article "VP9 Comes of Age, But Is it Right for Everyone?", I spoke with JWPlayer's lead compression engineer, Pooja Madan (bit.ly/vp9_age1). Madan designed and implemented JW's VP9 encoding facility, after spending many, many hours experimenting and testing. In the article, she shared her top four VP9 encoding takeaways, which were:

1. Use two-pass encoding; one pass does not perform well.

2. With two-pass encoding, generate the first pass log for the largest resolution and then reuse it for the other resolutions. VP9 handles this gracefully.

3. While VP9 allows much larger CRF values, we noticed that CRF <33 speeds up the encoding process considerably without significant losses in file size savings.

4. You must use the "tile-columns" parameter in the second pass. This provides multi-threaded encoding and decoding at minor costs to quality.

Our VP9 Encoding Ladder

Table 13-4 shows an encoding ladder for VP9 encodes. Basically, I used the same data rates as the HEVC ladder, and swapped out the VP9 settings for the x265.

	Width	Height	Frame Rate	Video Bitrate	Peak Bitrate	Key-frame	Threads	Speed	Tile-Columns	Auto-Alt-Ref	Lag-in-Frames	Audio Bitrate
270p	480	270	30	220,000	242,000	2	8	4/2	4	1	25	128,000
360p_l	640	360	30	400,000	440,000	2	8	4/2	4	1	25	128,000
360p_h	640	360	30	720,000	792,000	2	8	4/2	4	1	25	128,000
720p_l	1,280	720	30	1,000,000	1,100,000	2	8	4/2	4	1	25	128,000
720p_h	1,280	720	30	1,800,000	1,980,000	2	8	4/2	4	1	25	128,000
1080p_l	1,920	1,080	30	2,500,000	2,750,000	2	8	4/2	4	1	25	128,000
1080p_h	1,920	1,080	30	4,000,000	4,400,000	2	8	4/2	4	1	25	128,000
1440p	2,560	1,440	30	6,000,000	6,600,000	2	8	4/2	4	1	16	128,000
2160p	3,840	2,160	30	10,000,000	11,000,000	2	8	4/2	4	1	16	128,000

Table 13-4. Encoding ladder for VP9 encodes.

VP9 and FFmpeg

You can download a VP9 executable from www.webmproject.org, but it's easier to use FFmpeg.

1080p Conversion

Here's a batch file I used to create the 1080p and lower ladders from Table 13-4 from 1080p source. Note that I skipped the high-quality versions of the 1080p, 720p, and 360p files for simplicity. For completeness, the batch files contain the `auto-alt-ref` and `frame-parallel` switches in their default values that you can leave out if desired.

```
ffmpeg -y -i TOS_1080p.mov -c:v libvpx-vp9 -pass 1 -b:v 2500K -keyint_min
48 -g 48 -threads 8 -speed 4 -tile-columns 4 -auto-alt-ref 1 -lag-in-
frames 25 -frame-parallel 1 -f webm NUL && \

ffmpeg -i TOS_1080p.mov -c:v libvpx-vp9 -pass 2 -b:v 2500K -maxrate 2750K
-keyint_min 48 -g 48 -threads 8 -speed 2 -tile-columns 4 -auto-alt-ref
1 -lag-in-frames 25 -frame-parallel 1 -c:a libopus -b:a 128k -f webm
TOS_1080p_l.webm

ffmpeg -i TOS_1080p.mov -c:v libvpx-vp9 -pass 2 -s 1280x720 -b:v 1000K
-maxrate 1100K -keyint_min 48 -g 48 -threads 8 -speed 2 -tile-columns 4
-auto-alt-ref 1 -lag-in-frames 25 -frame-parallel 1 -c:a libopus -b:a 128k
-f webm TOS_720p_l.webm
```

```
ffmpeg -i TOS_1080p.mov -c:v libvpx-vp9 -pass 2 -s 640x360 -b:v 400K
-maxrate 440K -keyint_min 48 -g 48 -threads 8 -speed 2 -tile-columns 4
-auto-alt-ref 1 -lag-in-frames 25 -frame-parallel 1 -c:a libopus -b:a 128k
-f webm TOS_360p_l.webm

ffmpeg -i TOS_1080p.mov -c:v libvpx-vp9 -pass 2 -s 480x270 -b:v 220K
-maxrate 242K -keyint_min 48 -g 48 -threads 8 -speed 2 -tile-columns 4
-auto-alt-ref 1 -lag-in-frames 25 -frame-parallel 1 -c:a libopus -b:a 128k
-f webm TOS_270p.webm
```

Batch 13-1. Converting 1080p source to our VP9 encoding ladder 1080p rungs and below.

4K Scaling Exercise

I wanted to run the same scaling exercise with VP9 as I did with HEVC in the previous chapter, just to make sure that these scaling mechanisms worked for VP9 as well. To recount, the example inputs Tears of Steel at 3840x1714 resolution, with a display aspect ratio of 2.25:1 and outputs the encoding ladder shown in Table 12-3, save the high quality versions of the 1080p, 720p, and 360p files.

With the 4K and 2K files, I produced at full resolution (3840x2160, 2540x1440) with letterboxing, essentially using the command string shown in Batch 5-6. In all other resolutions I produced at full resolution while cropping out the excess pixels as shown in Batch 5-5.

Here are the commands.

```
ffmpeg -y -i TOS_4k.mov -c:v libvpx-vp9 -pass 1 -b:v 10000K -keyint_min 48
-g 48 -threads 8 -speed 4 -tile-columns 4 -auto-alt-ref 1 -lag-in-frames
25 -frame-parallel 1 -f webm NUL && \

ffmpeg -i TOS_4k.mov -c:v libvpx-vp9 -vf "scale=3840:2160:force_origi-
nal_aspect_ratio=decrease,pad=3840:2160:(ow-iw)/2:(oh-ih)/2" -pass 2 -b:v
10000K -maxrate 11000K -keyint_min 48 -g 48 -threads 8 -speed 2 -tile-
columns 4 -auto-alt-ref 1 -lag-in-frames 16 -frame-parallel 1 -c:a libopus
-b:a 128k -f webm TOS_4k.webm

ffmpeg -i TOS_4k.mov -c:v libvpx-vp9 -vf "scale=2560:1440:force_origi-
nal_aspect_ratio=decrease,pad=2560:1440:(ow-iw)/2:(oh-ih)/2" -pass 2 -b:v
6000K -maxrate 6600K -keyint_min 48 -g 48 -threads 8 -speed 2 -tile-col-
umns 4 -auto-alt-ref 1 -lag-in-frames 16 -frame-parallel 1 -c:a libopus
-b:a 128k -f webm TOS_2k.webm

ffmpeg -i TOS_4k.mov -c:v libvpx-vp9 -vf "scale=1920:1080:force_origi-
nal_aspect_ratio=increase,crop=1920:1080" -pass 2 -b:v 2500K -maxrate
2750K -keyint_min 48 -g 48 -threads 8 -speed 2 -tile-columns 4 -auto-alt-
ref 1 -lag-in-frames 25 -frame-parallel 1 -c:a libopus -b:a 128k -f webm
TOS_1080p_l.webm
```

```
ffmpeg -i TOS_4k.mov -c:v libvpx-vp9 -vf "scale=1280:720:force_origi-
nal_aspect_ratio=increase,crop=1280:720" -pass 2 -b:v 1000K -maxrate
1100K -keyint_min 48 -g 48 -threads 8 -speed 2 -tile-columns 4 -auto-alt-
ref 1 -lag-in-frames 25 -frame-parallel 1 -c:a libopus -b:a 128k -f webm
TOS_720p_1.webm

ffmpeg -i TOS_4k.mov -c:v libvpx-vp9 -vf "scale=640:360:force_original_as-
pect_ratio=increase,crop=640:360" -pass 2 -b:v 400K -maxrate 440K -keyint_
min 48 -g 48 -threads 8 -speed 2 -tile-columns 4 -auto-alt-ref 1 -lag-in-
frames 25 -frame-parallel 1 -c:a libopus -b:a 128k -f webm TOS_360p_1.webm

ffmpeg -i TOS_4k.mov -c:v libvpx-vp9 -vf "scale=480:270:force_original_as-
pect_ratio=increase,crop=480:270" -pass 2 -b:v 400K -maxrate 440K -keyint_
min 48 -g 48 -threads 8 -speed 2 -tile-columns 4 -auto-alt-ref 1 -lag-in-
frames 25 -frame-parallel 1 -c:a libopus -b:a 128k -f webm TOS_270p.webm
```

Batch 13-2. Converting 4K source to our VP9 encoding ladder with letterboxing and cropping.

So that's VP9. In the next chapter, we tackle a range of different skills that you'll find useful during your encoding and testing activities.

Chapter 14: Miscellaneous Operations

This chapter contains FFmpeg operations that are often critical to some task or another but didn't fit neatly into any of the previous chapters. Specifically, you will learn:

- what YUV and Y4M files are and how to produce them

- how to produce a PSNR score in FFmpeg

- how to concatenate multiple files into a single file without re-encoding.

Working With YUV/Y4M Files

Converting to and from YUV or Y4M formats is a frequent task for video producers, particularly those who use video quality measurement tools like the Moscow State University Video Quality Measurement Tool (VQMT), or those who want to encode to HEVC or VP9 with format-specific tools that only accept YUV/Y4M input. In these cases, you may have to use FFmpeg to convert to YUV/Y4M format to use these tools.

What's the difference between YUV and Y4M? YUV is a dumb file with sequential raw frames and a YUV extension. When working with these files, you may have to manually input format information like resolution, frame rate, or color space into the command string so the program knows what it's working with.

In contrast, when you create a Y4M file in FFmpeg, the program stores the resolution information and format metadata in the header. When you see Y4M, think metadata. Since the file contains this metadata, you can use a Y4M file as easily as an MP4 file; you don't have to specify resolution or format in either the GUI or command line.

This makes Y4M the easier format to use. As you'll see, creating a Y4M file is just a matter of specifying that file extension in the FFmpeg script. The only reason to use YUV is if the program you're working with won't input Y4M.

Converting with FFmpeg

Here's the command string to convert an MP4 file into Y4M output. As you've learned, since FFmpeg uses the native frame rate and resolution of the file unless told to do otherwise, you don't have to specify either of these to make the conversion.

```
ffmpeg -i TOS_1080p.mov -pix_fmt yuv420p -vsync 0 TOS_1080p.y4m
```

Batch 14-1. Converting a mov file to Y4M format.

`ffmpeg` calls the program.

`-i TOS_1080p.mov` identifies the input file.

`-pix_fmt yuv420p` identifies the file formats. The argument `-pix_fmt` tells FFmpeg to change the file format, while the `yuv420p` identifies the target format. I use `yuv420p` because it works with the Moscow University quality tool and other tools the require YUV/Y4M input, but it won't work in all cases, particularly when working with 10-bit formats.

`-vsync 0` tells FFmpeg to preserve the same video sync in the output file as in the input file. I'm not 100 percent sure that the `vsync` command is necessary—I inserted it in an attempt to eliminate some sync issues with a particular encoder, which it did, and it's never caused a problem with other encoders.

`TOS_1080p.y4m` the name of the target output file, and FFmpeg chooses the format based on the extension. I'm specifying Y4M for reasons discussed earlier. If you need a YUV file, just use that extension.

Note: YUV files can be very large, and can take a long time to process. This is one operation where working with SSD disks can really save you a bunch of time.

Scaling in FFmpeg

You learned how to scale for encoding back in Chapter 5. Often times when working with video quality measurement tools, you have to scale encoded files to larger resolutions to compare them to the source files. For example, if you produce a 360p file from 1080p source, and want to compute the PSNR value of the 360p file, you have to scale it back to 1080p to perform the analysis. Here's the command line for doing that, with the new arguments in bold.

```
ffmpeg -i  TOS_360p.MP4 -pix_fmt yuv420p -vsync 0 -s 1920x1080
-sws_flags lanczos TOS_1080p.y4m
```

Batch 14-2. Converting and scaling a file to Y4M format.

Here's an explanation of the new switches in the command line argument.

`-sws_flags lanczos` tells FFmpeg to use the Lanczos filter to perform the scaling.

I used the Lanczos scaling method after finding a white paper from graphics card vendor NVIDIA that stated this was a primary method used on the company's graphics cards. Since I was trying to simulate the quality perceived when a graphics card scaled video, this seemed appropriate. Lanczos is not the default method for FFmpeg, although the documents don't appear to specify what the default method is (see bit.ly/ff_scale). I compared the quality of a file produced with

Lanczos against one produced without specifying a method (which obviously used the default method), and the default method rated slightly higher. So, when scaling to simulate a graphics card, use Lanczos; when trying for top quality, don't specify and use the default (although you can Google and find plenty of tests that disagree with my results).

Computing PSNR with FFmpeg

One of the video quality metrics that FFmpeg can provide is the Peak Signal-to-Noise Ratio (PSNR) you learned about in the Introduction. There are two elements to computing PSNR; first computing it, second recording it into a log file, because if you don't create the log file, the PSNR value doesn't get saved.

```
ffmpeg.exe -i TOS_1080p.mov -c:v libx264 -b:v 2500k -psnr -report
TOS_1080p_2500.mp4
```

Batch 14-3. Converting a file and computing the PSNR value.

Here's an explanation for the new commands used in this string.

`-psnr` Tells FFmpeg to compute PSNR.

`-report` Tells FFmpeg to store a log file, which it will name according to the date and time (`ffmpeg-20170417-193149.log`).

If you open the log file and scan to the bottom, you'll see PSNR values for Y, U, and V (Figure 14-1). We'll exclusively use the Y value in this book, so that's the one you want.

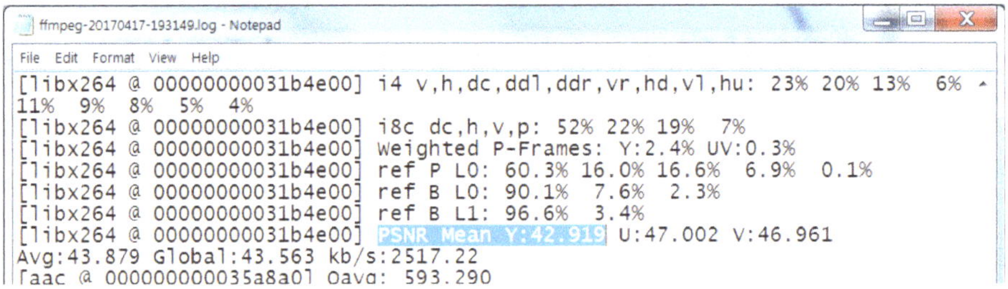


```
[libx264 @ 00000000031b4e00] i4 v,h,dc,ddl,ddr,vr,hd,vl,hu: 23% 20% 13%  6%
11%  9%  8%  5%  4%
[libx264 @ 00000000031b4e00] i8c dc,h,v,p: 52% 22% 19%  7%
[libx264 @ 00000000031b4e00] Weighted P-Frames: Y:2.4% UV:0.3%
[libx264 @ 00000000031b4e00] ref P L0: 60.3% 16.0% 16.6%  6.9%  0.1%
[libx264 @ 00000000031b4e00] ref B L0: 90.1%  7.6%  2.3%
[libx264 @ 00000000031b4e00] ref B L1: 96.6%  3.4%
[libx264 @ 00000000031b4e00] PSNR Mean Y:42.919 U:47.002 V:46.961
Avg:43.879 Global:43.563 kb/s:2517.22
[aac @ 000000000035a8a0] Qavg: 593.290
```

Figure 14-1. Here's the PSNR Mean for the Y value.

For those who care about such things, FFmpeg's numbers do not exactly match those reported by the Moscow State University VQMT Tool. In the nine cases that I checked, FFmpeg proved about 1.8% higher, averaging 38.6 compared to 37.9 for VQMT, which isn't significant. Scores were consistently higher, ranging from about 1.75% to 2.08%, rather than all over the map. If I didn't have VQMT to use, I would definitely fall back on FFmpeg.

Concatenating Multiple Files

Use this procedure to concatenate multiple files without reencoding. The first step is to create a list of all the files. Figure 14-2 shows the list I created for this operation.

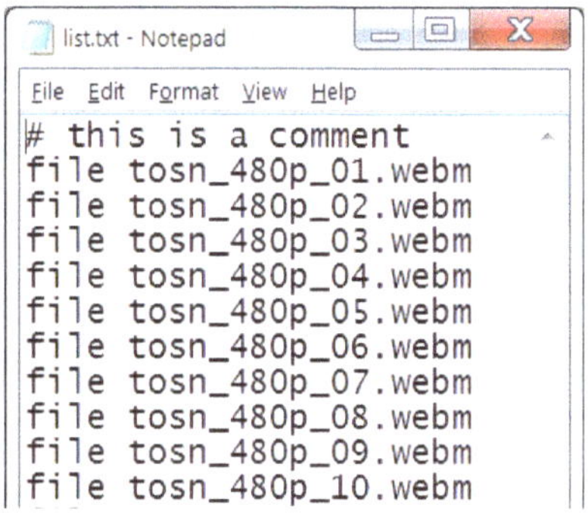

Figure 14-2. The list of files to concatenate.

The list should start with a line other than the start of the list; hence the `# this is a comment`. You must include the word `file` in the description, and you can include a path to the file if desired. I saved this list in the same folder as the files, and here's the FFmpeg command string to make it work.

```
ffmpeg -f concat -i list.txt -c copy combined.webm
```

Batch 14-4. The command to concatenate the files and produce test.webm.

Here's an explanation for the new commands used in this string.

- `-f concat` - Tells FFmpeg to concatenate the files.

- `-i list.txt` - Identifies the input file.

- `-c copy` - copies the audio and video codecs.

- `combined.webm` - Identifies the output file.

Note that this command does not work on all formats, most notably MP4. However, if you convert the MP4 files to MPEG-2 transport streams (Batch 3-4) first, you should be able to concatenate them, and then convert the longer file to MP4.

I'm sure there are tons more operations worthy of inclusion in this chapter, but that's all we have time for. I hope you found this book helpful. Please send any errors or suggestions to me at ffmpeg@streaminglearningcenter.com.

Until next time, take care.

Index

www.ingramcontent.com/pod-product-compliance
Lightning Source LLC
Chambersburg PA
CBHW041007050326
40690CB00029B/5289